ALSO BY NICHOLAS WADE

*A World Beyond Healing: The Prologue
and Aftermath of Nuclear War*

Betrayers of the Truth (with William Broad)

*The Nobel Duel: Two Scientists' 21-Year Race to Win the
World's Most Coveted Research Prize*

The Ultimate Experiment

Life Script

How the Human Genome Discoveries
Will Transform Medicine
and Enhance Your Health

NICHOLAS WADE

A TOUCHSTONE BOOK
Published by Simon & Schuster
New York   London
Toronto   Sydney   Singapore

TOUCHSTONE
Rockefeller Center
1230 Avenue of the Americas
New York, NY 10020

First Touchstone Edition 2002

For information about special discounts for bulk purchases,
please contact Simon & Schuster Special Sales at 1-800-456-6798
or business@simonandschuster.com

Designed by DEIRDRE C. AMTHOR

Manufactured in the United States of America

1   3   5   7   9   10   8   6   4   2

The Library of Congress has cataloged the
Simon & Schuster edition as follows:
Wade, Nicholas.
Life script: how the human genome discoveries will transform
medicine and enhance your health / Nicholas Wade.
p.   cm.
Includes bibliographical references and index.
1. Medical genetics—Popular works.   2. Human genome—
Popular works.   3. Human Genome Project—Popular works.
I. Title.
RB155.W326   2001
616'.042—dc21   2001042627

ISBN 0-7432-1605-9
0-7432-2318-7 (Pbk)

Contents

Preface

The human genome embodies the genetic program and parts list needed to make, operate and maintain a person.

Possession of this human instruction manual, first decoded in June 2000, marks the beginning of a new era of medicine. It provides the basis on which to understand the human body almost as fully and precisely as an engineer understands a machine. From that understanding, physicians can hope to develop new ways to fix the human machine and in time to correct most—perhaps almost all—of its defects.

The purpose of this book is to describe the dawning of the genomic revolution and the ways in which the new knowledge is likely to transform medicine and human health. Though the future of a technology is hard to predict, there are certain obvious paths to pursue now that the human instruction manual is in hand. And though this revolution is being conducted in the name of medicine, it will not necessarily stop at its implied goal, the attainment of perfect health. The power to reshape the human clay has no clear limits. How far should we go in enhancing qualities other than health, such as physique or intelligence? If cures are developed for the major degenerative diseases of old age, how greatly can life span be extended without undermining social institutions that are designed around an orderly cycle of birth, procreation and death? And if the new genetic medicine works so well, wouldn't it be better to apply it in the germline, giving everyone the genetic endowment for a long and healthy life?

Such questions are futuristic, yet the genomic revolution is unfolding so fast that almost no speculation about the physical basis of human existence is premature. New words such as genomics and proteomics denote plans to accelerate biology by studying genes and proteins on a genome-wide basis instead of just one at a time, as has been the practice hitherto. With the genome sequence in hand, biologists can for the first time hope to understand that maestro of the genome, the human cell, which at every instant is reading off hundreds or thousands of genes on a scale of activity that has until now been far too complex to track. And the cell, the unit of which the body is built, is the key to understanding the whole human organism and its operation in both health and disease.

Both at universities and in the private sector, vigorous efforts are under way to gather the genome's first fruits. The race to decode the human genome sequence has left in place two powerful competitors whose rivalry—much in the public interest—seems likely to continue long into the post-genome era. One is the international consortium of academic centers that began sequencing the human genome in October 1990 in an endeavor known as the Human Genome Project. They and other academic biologists, now increasingly networked, are bending every effort to interpret the human genome sequence. Meanwhile Celera Genomics, the private company that leapt in to challenge the consortium at the final lap of the Human Genome Project, has continued to spearhead the race to interpret the genome. It is sequencing the genomes of other animals to help biologists better understand the construction of the human genome. It is building its own database, centered on its human genome sequence, in direct competition with that developed by its academic rivals. These genome databases are new frameworks for organizing all biological and medical knowledge, much as the periodic table of elements organizes all of chemistry. They are also a powerful heuristic tool for interpreting the knowledge they contain.

Other genomics companies, such as Human Genome Sciences, aim to translate genomic data directly into new drugs based largely on human proteins. The traditional pharmaceutical companies are built around easily synthesized, small molecule chemicals that act on fewer than five hundred protein targets, which is all that pharmacologists had managed to discover in the pre-genome era. The human genome sequence makes available the full set of human proteins, of which there appear to be at least thirty thousand. The pharmaceutical industry is thus poised to undergo a double revolution. A host of new small molecule drugs can be designed to target the new proteins. And a new wave of protein-based drugs is about to emerge, following in the steps of inventions such as Amgen's erythropoietin for stimulating blood formation and Immunex's Enbrel for rheumatoid arthritis.

Besides new drugs, medicine seems likely to be transformed in the near future by a new wave of genome-derived diagnostic tests. DNA molecules can be attached to small squares of glass that, by analogy with computer circuits, are called microarrays or gene chips. Most human genes exist in several forms that differ very slightly in their DNA sequence, and some of these variant genes cause or contribute to disease. Gene chips can be designed that will detect these variant genes and thus diagnose a patient's vulnerability to any disease whose underlying gene variants are known.

A new phrase, individualized medicine, has been coined to describe the concept of treating patients in light of a genome scan that predicts the diseases they are likely to develop in the course of their lifetime. It already seems technically feasible to construct gene chips that would test for the presence of all the common variants of every human gene. These gene chips would in effect sequence a person's entire genome simply by recording the genome's principal points of difference with the consensus human DNA sequence. Such a wealth of genetic information requires the most careful handling and confidentiality, and in practice genome scans are likely to be used at first in a very constrained fashion; there is little point in testing people for diseases for which there is no treatment.

A prominent aspect of individualized medicine is the idea of marketing certain drugs along with a genetic test that would screen out patients who are likely to have adverse reactions. The tests might also show which of several equivalent drugs would best suit a particular patient. Pharmacogenomics is the new name for matching patients and drugs, a procedure that may reduce the vast burden of adverse drug reactions and rescue many drugs that otherwise would be too dangerous to use.

Knowledge from the genome seems likely soon to meld with new knowledge about human cells to create a novel kind of healing that some are calling cell therapy or regenerative medicine. The audacious notion of regenerative medicine is to heal the body with tools from its own repair kit, growing new tissues and organs to replace those weakened by age or disease. It seems that many of the body's components possess a reservoir of self-renewing cells, known as stem cells, which are used to repair and maintain the tissue. Unfortunately, perhaps to curb the risk of cancer, nature has strictly limited the growth and vigor of human stem cells. Hence people do not grow new hearts or livers when needed as a lizard grows a new tail, even though all the design information must still be present in the genome. The hope of stem cell biologists is to unlock the inherent plasticity of human cells and to mold them, perhaps with the help of the natural control signals that can be identified through the genome, into replacement parts for diseased and aging tissues. Present day medicine often enables patients to survive with damaged tissues but does not aspire to cure them. The would-be practitioners of regenerative medicine intend to grow new and youthful tissues, probably from a patient's own stem cells, and to restore the person to full and vigorous health.

The cascade of innovation generated by the human genome sequence seems likely to bring many blessings. But it will in time raise some interesting dilemmas. With the human instruction manual in hand, as well as those of pathogenic viruses and bacteria, will any identifiable barrier remain to curing all human disease? If and when all disease is conquered and Pandora's box at last hammered shut, a great goal of medicine will have been attained. People will live out their natural life span in full health. But genomic knowledge may then open a new door, one that will allow life span itself to be extended.

Researchers have already learned how to stretch the life of laboratory organisms such as roundworms and fruit flies by manipulating certain genes. They have "immortalized" human cells, meaning that the cells can be made to grow in the laboratory indefinitely far beyond the point, measured in terms of cell divisions, at which they would usually lapse into senescence.

Though the familiar facts of death and disease make us think of our bodies as inherently frail and perishable, the living cells of which they are made are the hardiest of survivors. Evolution has made the cell serve a wide variety of life spans depending on each organism's needs. Some mayflies live only a few hours on reaching adulthood. But the bristlecone pine endures for five thousand years. Genomic knowledge may show us not only how to correct the degenerative diseases of age from which evolution has neglected to protect us, but also how to unlock mechanisms that determine life span. A long and healthy life is a true blessing, but ultra-long life could challenge many human rites and customs.

Another delicate issue that the genomic revolution will raise is that of altering the human germline. If genetic medicine is as successful as hoped, there could be a logical case for avoiding a lifetime of expensive medical care in favor of building genetic improvements directly into the human germline. The technology for adding large numbers of genes to the human germline does not exist at present but there may be no insuperable obstacle to doing so should the demand arise and receive social sanction.

Altering the human genome to suppress disease-causing gene variants would promote health, but parents might then wish to improve other attributes such as physique or intelligence. How far, if at all, could these qualities be enhanced without changing human nature?

Despite the many serious objections to manipulating the human genome, it is easy to envisage the trends that might favor this momentous step. The rapidly increasing knowledge of the genome will transform medicine but is likely also to reveal the human genome as a work in progress, rough hewn by evolution and shot through with oversights that are the cause of infinite human misery. Physicians may urge the case for removing these defects. As pharmacogenomic tests become routine, patients will become increasingly aware of their genetic vulnerabilities and perhaps concerned not to bequeath them to their children. And as biologists learn how to repair the human fabric, they may see how to realize evolution's design more fully, by taking shortcuts to goals that evolution would reach more circuitously over thousands of years.

The pages that follow, which draw on reporting done for articles in *The New York Times,* aim to sketch out the events that led to the genome era, the likely first applications of the human genome sequence to health, and the prospects of more radical innovations such as regenerative medicine and the extension of natural life span. No conceivable body of knowledge can rival the value or fascination of the human genome. I hope readers will enjoy this attempt to peer over biologists' shoulders as in wonder and dawning comprehension they turn the first pages of the human life script.

1
The Most Wondrous Map

"Nearly two centuries ago, in this room, on this floor, Thomas Jefferson and a trusted aide spread out a magnificent map—a map Jefferson had long prayed he would get to see in his lifetime. The aide was Meriwether Lewis, and the map was the product of his courageous expedition across the American frontier, all the way to the Pacific."

So said President Bill Clinton, speaking in the East Room of the White House to an audience of scientists and government officials and, by teleconference, with Prime Minister Tony Blair in London. The occasion, on the morning of Monday, June 26, 2000, marked the completion of the Human Genome Project or rather, since in truth the genome was almost but not entirely finished, "the completion of the first survey of the entire human genome."[1]

In the audience were the members of the two rival teams who had battled each other to the finish line. The race for scientific distinction is often fierce, and there are few greater scientific prizes than the human genome. This race was fueled by political differences as well as hope for glory. A consortium of academic biologists in the United States and Britain had embarked on decoding the human genome with the intent of making it a communal good, freely available to the scientific and medical communities. Though they were not oblivious to the personal rewards of success, their goal was the disinterested creation of public

wealth, in the belief that medical advance would be speediest if all biologists had full and free access to the human genome sequence.

But in May 1998, almost eight years after the public consortium had
begun to lay the technical groundwork and was within sight of success,
a commercial enterprise headed by J. Craig Venter jumped into the fray
with the goal of decoding the human genome as a profit-making venture. Venter, formerly an academic biologist like his rivals, believed that
speed too was a public good and that with centralized management and
a different strategy he could sequence the human genome much faster
than the consortium seemed likely to do.

With such different motives and such high stakes, the race for the
genome was especially intense. Each team confidently predicted that
the other's strategy would fail, while working round the clock to ensure
its own success. As the finishing post neared, it became clear that each
had achieved enough of its own goals to assert a plausible claim of victory. Venter's company, Celera, had the edge on the scientific front,
since his genome sequence was more complete than the public consortium's. But the consortium had achieved its political goal: its version of
the human genome, containing almost all human genes, was publicly
available for the free use of scientists around the world.

In June 2000, the two sides were persuaded at the last moment that
each stood to gain more from a decorous declaration of joint victory
than from rival assertions of victory, spiced with mutual derogation of
the other's achievement.

And so it was that Clinton, in his speechwriter's metaphor, came to
compare the decoding of the human genome with Lewis's historic map
of an America bounded by the Pacific and Atlantic oceans.

"Today," the president said, "the world is joining us here in the East
Room to behold a map of even greater significance. We are here to cel-
ebrate the completion of the first survey of the entire human genome.
Without a doubt, this is the most important, most wondrous map ever
produced by humankind."

Shifting analogies, the president went on to compare the sequencing
of the genomic script to "learning the language in which God created
life. . . . With this profound new knowledge, humankind is on the verge

of gaining immense, new power to heal. Genome science will have a real impact on all our lives—and even more, on the lives of our children. It will revolutionize the diagnosis, prevention, and treatment of most, if not all, human diseases. . . . In fact, it is now conceivable that our children's children will know the term 'cancer' only as a constellation of stars."

The scene then shifted to a large television monitor, where from London Prime Minister Tony Blair praised the "huge role" the United States had played in bringing the Human Genome Project to fruition. "Huge role" does not mean "leading role" or "central role" or even "essential role." The backhanded compliment was Blair's way of signaling that the project was British in its roots, though implemented with American money. To underscore the point, Blair had seated with him Fred Sanger, the biologist after whom Britain's DNA sequencing center was named. Sanger, a genial, unassuming fellow who has made few public utterances, invented a highly ingenious method of decoding the sequence of chemical units in DNA. Almost every problem the two teams of sequencers had grappled with as they completed the human genome, Sanger had wrestled with twenty years before. The only biologist to win two Nobel Prizes—one for a method of sequencing proteins, one for his DNA method—Sanger launched the field of genomics in 1977 by sequencing the genome of a small virus 5,375 units in length. But without advanced computers, automation, and a method of amplifying DNA not invented until 1985, Sanger had been unable to take his tour de force further.

The two teams that decoded the human genome had depended on Sanger's method, though with its chemistry altered so that it could be performed by machine instead of manually. The Human Genome Project started in earnest when John Sulston at the Sanger Centre, with his American partner Robert Waterston, began to test the new sequencing machines on a pilot project, the genome of the roundworm, a microscopic animal much studied in laboratories. The center's work was supported by the Wellcome Trust of London, the world's largest medical philanthropy.

"For let us be in no doubt about what we are witnessing today," Blair continued, "a revolution in medical science whose implications far surpass even the discovery of antibiotics, the first great technological triumph of the twenty-first century." Today's announcement, the prime minister said, "opens the way for massive advances in the treatment of cancer and hereditary diseases, and that is only the beginning."

Clinton then turned to Francis Collins, director of the National Human Genome Research Institute and the leader of the academic consortium. Collins, a born again Christian, said, "It is humbling for me and awe-inspiring to realize that we have caught the first glimpse of our own instruction book, previously known only to God."

Though Clinton and Blair had already praised Venter, Collins in introducing him heaped further laurels on his head, twice calling his strategy innovative and referring to his "landmark achievement" in assembling his own genome sequence.

With a build-up of such authority, Venter saw no particular need to minimize the significance of his achievement. "Today, June 26 in the year 2000," he began, "marks a historic point in the 100,000-year record of humanity. We're announcing today for the first time our species can read the chemical letters of its genetic code."

The human genome sequence, Venter said, "represents a new starting point for science and medicine, with potential impact on every disease." He suggested that the genome might even generate quick treatments for cancer, saying, "There's at least the potential to reduce the number of cancer deaths to zero during our lifetimes."

Could anything, even the human genome sequence, live up to such intense billing? A turning point in history, a divine revelation, a cure for cancer, all rolled into one?

Some critics were quick to suggest that disease-causing gene variants might be very hard to find, despite having the genome sequence in hand, and that even if they were, few people would want to be tested unless treatments were available too. "The new genetics will not revolutionize the way in which common diseases are identified or prevented," two skeptics wrote in *The New England Journal of Medicine*.[2]

The harshest verdict came from William Haseltine, Venter's former partner and chief executive of Human Genome Sciences. "Most of us in the pharmaceutical industry would agree that the draft sequence basically is a non-event in our world," he said of the genome sequence to *The Wall Street Journal* a month after the White House announcement.[3] Elsewhere he described the announcement as "a symbolic moment, like sending men to the moon symbolized our intent to use space, like Admiral Peary's journey to the North Pole symbolized the intent to explore the Arctic."[4]

Haseltine's point was not that the genome sequence was as dubious as Peary's claim to have reached the pole but that he himself had chosen a much faster method of exploiting its information, by extracting ready-made transcripts of genes from the cell instead of ferreting for the genes themselves in the genome sequence. Many biologists, however, would agree in principle with the optimistic assessments offered at the White House, although probably few would offer any kind of timetable for curing cancer. For several reasons, the sequencing of the human genome does indeed mark a milestone in the history of science and medicine, perhaps in human history too.

Just as Meriwether Lewis's map showed the finiteness of America, vast as it was, the genome for the first time places limits on human biology. The working of the human body is no longer a boundless mystery. There's still enigma enough, but the genome sequence defines the extent of the problems to be solved and bolsters biologists' confidence that they will eventually be soluble. That means that the genetic roots of disease will at least be understood in exquisite detail. And the usual long interval between laboratory understanding and practical treatments may shrink as the genome quickens the pace of every aspect of biology.

The promise of the genome is almost unimaginably broad. Every disease is caused or in some measure influenced by a person's genes. This is true even of infectious diseases: their prime cause may be a bacterium or virus but people vary widely in their susceptibility, depending on their genetic make-up. The major degenerative diseases of old age, such as cancer, heart disease, diabetes and Alzheimer's, are strongly influenced by genes that predispose a person toward them.

Possession of the human genome sequence means that the genes that contribute to these diseases—they are variant versions of normal genes, not special genes whose only role is to wreak havoc—can be tracked down and studied. In the pre-genome era biologists had some limited successes in identifying variant genes, but the genome sequence is expected to make gene hunting much faster and more systematic.

Discovery of a variant gene by itself does nothing for a patient. But it makes an excellent start to understanding the exact mechanism of a disease. Researchers can examine the protein product that is specified by the variant gene, identify the protein's usual role in the cell and how it differs because of the variation in the gene, and figure out how the aberrant protein causes or contributes to the symptoms of the disease.

The present proposals for developing medical treatments from genomic knowledge may or may not be successful, as is always the case with research, but there is a widespread view among biologists that a new era has begun, one that incorporates all the single gene biology of the past and enables living cells to be studied for the first time in their totality, on a genome-wide basis.

The genome is the infrastructure that unites all the branches of biology; in doing so, it promises to accelerate dramatically the pace of advance in medicine. Most basic biology, such as study of how genes and cells work, is done in laboratory organisms such as the mouse, the fruit fly, and the roundworm, and then confirmed in human cells. The knowledge gained from these animals can be related much more quickly to humans now that the genomes of these species have been sequenced and can be compared directly. With a few keystrokes, a researcher can now ascertain whether a new gene found in the mouse or fruit fly or roundworm has counterparts in people. The mouse in particular, a fellow mammal, has turned out to be surprisingly similar to people on a genomic level and all the more helpful for that reason in interpreting the human genome.

The human genome sequence will prove invaluable in another way: it provides the basis for explaining and integrating all knowledge about the human body. For example, biologists are already beginning to understand the anatomy and physiology of the cochlea—the organ of hearing—in terms of the genes that that make each of its components.[5] This anatomical dissection at the genetic level has been possible through studying the genetic causes of deafness. The ear is a delicate mechanism, with such fine tolerances that a mutation in almost any of the genes involved in building it causes loss of hearing. Through these mutations, the genes themselves have been discovered. It is easy to envisage the time when all the body's other organs will be described in terms of the genes that specify their development in the embryo and their function and maintenance throughout life. The genome will become a unifying explanation for all of human biology and medicine.

The genome, in other words, is more than just a sequence. It is a means of reorienting biology and medicine and of accelerating the pace of discovery in all the biomedical sciences.

Beyond its practical importance, the genome bears on the nature of human life. It does not say what life is nor determine the life of any individual. It may not reach to the deepest mysteries of human existence. But it defines with great precision every component of every human cell, and these components specify with great exactness the steps from an egg to a newborn. The rules for the architecture of the brain are surely implicit in the genome, even though the genome defines only genes. The infinite varieties of human thought and behavior are too complex to be determined by the finite information embodied in the genome. But behavioral information can be inscribed in the genome, as is already known from the discovery of genes that determine feeding behavior in the roundworm and courtship rituals in the fruit fly. It is certainly possible, even likely, that many general rules of human behavior are implicit in the genes.

Humans have far fewer genes than was generally expected to be the case. The fruit fly has almost 14,000 genes, the roundworm 19,000, and people only 30,000 or so. The enormous increase in complexity between a worm and a person is obviously not reflected in the only slightly greater number of genes and must lie in other factors, such as the more complex nature of human proteins, which allows more inter-actions between them.

Biologists, as they consider the genome's impact on their work, are also moved by the prospect of doing science in new ways. Analysis of the genome has to be done on computer, a circumstance that is forcing the digitization of other aspects of biology. There used to be two kinds of biological experiment: in vitro and in vivo (in glassware and in live animals). There is now a third: in silico, meaning biology that is done by querying genome data in distant data banks. "The instructions for as-sembling every organism on the planet—slugs and sequoias, peacocks and parasites, whales and wasps—are all specified in DNA sequences that can be translated into digital information and stored in a computer for analysis," two biologists wrote recently. "As a consequence of this revolution, biology in the twenty-first century is rapidly becoming an information science. Hypotheses will arise as often in silico as in vitro."[6]

Some observers expect the pace of biological research to accelerate because so many of the world's biologists are becoming networked by the need to consult genomic databases. "Networking biology started in the 1990s and we think will accelerate to the point where genomic and protein information will become available to all scientists; that will be a vast change coming over the research community in the next decade," says Randal Scott, chairman of Incyte.[7]

The genome is important for another reason: it may reflect the be-ginning of an intellectual turning point in biological research. Molecu-lar biology's successes have come from a rigorous reductionism—the approach of ignoring the whole organism, however interesting, in favor of reducing it to components that can be defined and understood. Though critics have accused scientists of ignoring all that is important about life, it was a rational and essential program to begin with the as-pects of life that could be analyzed: genes, proteins and cells.

The human genome sequence is in a sense the ultimate triumph of reductionism—a list that represents the entire parts list for constructing and operating a person. But to understand the genome biologists must now swing into reverse and figure out how the parts work together. With the help of gene chips that monitor the simultaneous activity of thousands of genes in a cell, it will for the first time be possible to study a complete cell at work. Biology may move into a phase of synthesis as researchers try to reconstruct the pathways between the genes in the genome and the living organism.

Perhaps the greatest impact of the human genome is that it lays the basis for a systematic, comprehensive approach to medical discovery. For almost any disease, researchers can in principle start by searching for its genetic roots in the human genome. Even with diseases caused by viruses or bacteria, people vary widely in their resistance to such diseases, and the basis for such resistance is genetically determined. Since 1995, biologists have already laid bare the genomes of most pathogens, so that a systematic approach can be developed to infectious diseases too.

The comprehensive strategy that the human genome makes possible is a far cry from the traditional method of medical discovery with its heavy dependence on luck and inspiration.[8] In the age of genomics, researchers should be less dependent on happy accidents of the kind that accompanied Alexander Fleming's discovery of penicillin. As it happens, the power of antibiotics is now waning as more and more species of bacteria acquire the genes to resist them. But instead of waiting for another Fleming, biologists can scan the genome of a pathogen, searching for weak points in its defenses where a drug or vaccine might be brought to bear. For any human disease the genome provides a powerful new starting point and a means of accelerating research leads from elsewhere.

The superlatives that flowed at the White House announcement were not without tangible basis. It is hard to think of any other body of knowledge that rivals the human genome in significance. It was a remarkable scientific triumph that the full sequence of the human genome should emerge less than fifty years after DNA had first been identified as the repository of hereditary information. Indeed, the goal of the public consortium is still to produce its final version of the genome in 2003, on the fiftieth anniversary of James Watson and Francis Crick's 1953 discovery.

Watson, who had also played a central role in launching the public consortium's project, was present in the audience. It was hard to tell how pleased he was with the turn of events. He had overcome the resistance of many leading biologists to win government support for the project and served as its first director at the National Institutes of Health. Yet for him the day's triumph was bittersweet. The genome was complete, or almost so, but Venter, a man with whom Watson had had serious differences, had grabbed a large share of what would otherwise have been the Watson consortium's credit for this most glorious of scientific achievements.

It was perhaps little solace that the president drew attention to Watson's presence, noting his involvement in the discovery of both DNA and the genome sequence. After the ceremony, reporters surrounded Watson, seeking his reaction to the day's event. But Watson, often candid to a fault, had strangely little to say. The genome, his silence implied, spoke for itself, and his own emotions about the race he had not quite won were not for others' ears.

2
The Race for the Human Genome

James D. Watson's career has spanned a remarkable period in the history of biology, from the discovery of the structure of DNA to the decoding of the human genome. Though these events were forty-seven years apart, Watson played a central role in both. But the genome sequence, which might have capped his career as an unalloyed triumph, was one he was compelled to share with an opponent whose style and character were not so dissimilar from his own.

Like Watson, J. Craig Venter is sharply focused on goals and pragmatic in achieving them. Both men are candid to a fault, often sharing thoughts or scathing criticisms that others would bite back. Yet both, despite their penchant for undiplomatic frankness, are successful organizers and builders of coalitions.

The contest between Venter and the consortium Watson put together was a political as well as a personal one. The consortium envisaged the genome sequence as the basic infrastructure for future biology, a public road that every researcher should have a right to travel. Venter believed his centralized attack on the genome would produce a better product sooner and that the small fee Celera would be obliged to charge academic departments was no different from a highway toll or a journal subscription. Watson's idea of distributing the genome effort around several academic centers may have been politically adroit but made no technical sense once a standard sequencing methodology had emerged. Sequencing the human genome in a dozen different academic centers, with different procedures and varying standards of quality control, was in Venter's view a recipe for delay, if not outright disaster.

That Watson and Venter found themselves on opposite sides of the issue was a matter more of circumstances than of conviction. Watson had had to forge his consortium with money from governments and the Wellcome Trust of London before serious commercial backing became available. Venter spent the first part of his career in the public sector and might have continued to do so if Watson and other decision makers in the consortium had provided the support he sought from them. Only after the consortium had declined to embrace his ideas did Venter accept the private money that came chasing after them.

In the end it was hard for observers to know which contender to root for, since each embodied different virtues. The public consortium put together by Watson and his successor, Francis Collins, was engaged in an indisputably unselfish enterprise. By custom, academic scientists keep their findings to themselves for a few months so that they can gain credit on publication for harvesting the first fruits of their discoveries. But the consortium's scientists agreed to make their human genome data available each night on public databases, taking no credit other than for producing it.

Besides being disinterested, the consortium, by Watson's deliberate design, was international. "It wouldn't be good if the Americans owned the genome," he said.[1] He vigorously campaigned for Germany and Japan to join the project. As a result of his efforts, the consortium came to include laboratories in Britain, France, Germany, China and Japan. Although the vast bulk of the genome sequencing was accomplished by the centers in Britain and the United States, with Britain's Wellcome Trust paying for 30 percent of the sequence, the consortium's composition and openness refuted any charge that the United States was trying to appropriate the common heritage of humankind.

The consortium succeeded in putting an invaluable resource at researchers' disposal. During the years from 1998 to the end of 2000, when the bulk of the human genome sequence was being decoded, biologists were able to tap into GenBank, the DNA database run by the National Institutes of Health, and search for human genes of interest, whereas Celera's human genome was not accessible until early 2001.

Moreover, the consortium was a technical success. Despite its highly unwieldy, decentralized structure, it surmounted all obstacles and was on track to completing the human genome on time and within budget, a rare achievement for a government project.

But there was also outstanding merit in Venter's approach. By tackling the genome in a single facility, he reaped the advantages of scale and uniform quality. Showing excellent technical judgment, he chose a faster strategy for decoding the genome. He put extra effort into computer analysis to offset his strategy's higher risks, confounding his critics in the consortium who had rashly predicted that the strategy was bound to fail. Once his facility was up and running and his strategy proved by a test run on the genome of the *Drosophila* fruit fly, he sequenced the human genome in the amazingly brief period of ten months.

"Every time you are in Watson's presence you realize you are in contact with a man who has changed the course of human history and who will be remembered long after you have turned into dust," says Phillip Sharp, a leading biologist at the Massachusetts Institute of Technology.[2] Like many others, he ranks Watson and Crick in the same circle as Charles Darwin and Gregor Mendel as pivotal figures in the history of biology. But whereas Crick went on to dominate the early years of biological research that laid the foundations of molecular biology, Watson soon became a scientific administrator, building world class biology departments at Harvard and later at the Cold Spring Harbor Laboratory on Long Island, New York.

No one in Watson's presence could imagine they were talking to an ordinary individual. His wayward wisps of gray hair suggest an obliviousness to routine concerns. As he talks he seems to stare past a visitor with his ice-blue eyes. Then suddenly his eyebrows arch, his face a picture of astonishment, as if he himself were dazzled with the clarity of the point he has just made.

Clarity of vision has been a hallmark of Watson's career. He discovered the structure of DNA at the age of twenty-five because he understood ahead of others that it must hold the answer to how the hereditary information is stored and duplicated. Crick figured out that the DNA molecule must consist of a double helix, two intertwined spiral chains, but it was Watson who first spotted how a pair of bases, one from each chain, must cross-link to form each step of the spiral staircase.

It was the same clarity of vision, some four decades later, that led Watson to see the necessity of sequencing the human genome. "It was so obvious you had to do it," he said later. To him maybe, but many biologists strenuously opposed the idea, first formally proposed by Robert L. Sinsheimer of the University of California at Santa Cruz, at a meeting he convened in 1985.[3] The Department of Energy first took up the idea at its Los Alamos National Laboratory. The lab specialized in nuclear weapons design; its rationale for getting involved in the genome was that its radiation biologists needed to understand the genetic effects of radiation.

Watson believed strongly that if the human genome were to be sequenced, it should be done by academic biologists and therefore under the sponsorship of their federal government patron, the National Institutes of Health, or NIH. Many biologists opposed the project for the same reason: they feared that a big genome program by their patron agency would inevitably diminish the funds available for their own grants.

Molecular biologists were also intellectually opposed to projects designed just to gather data. Good science, in their view, is deductive, characterized by elegant experiments that test a specific hypothesis about how nature might work. Inductive science—the mere amassing of scientific facts—is more like stamp collecting than proper science in their view. "I think biology has been most effective as a cottage industry and should remain so as long as it can," said David Baltimore, then president of Rockefeller University and now president of Caltech.[4] Robert Weinberg, a cancer geneticist at the Whitehead Institute in Boston, said that sequencing the 97 percent of DNA that does not code for genes would be wasted effort because "I believe it will turn out to be biologically meaningless."[5]

Watson urged on his colleagues the need for an NIH role in deciphering the genome and lobbied Congress to approve extra funds. Though he did not seek the role, it soon became clear that Watson was the most acceptable choice to lead the project, and in October 1988 he was appointed the first director of the Office of Genome Research within the NIH, a post he held until his resignation in 1992.

Watson quickly laid out the principal elements in the public consortium's strategy. His decision to site genome institutes at several universities around the country instead of in some central location deftly defused academic opposition to government-directed research. It also helped spread the money to various constituencies, always a prudent disposition for projects that require annual appropriations from Congress.

Foreseeing the need to interpret the human genome once it was completed, Watson decreed that the genomes of several other organisms should be sequenced as part of the Human Genome Project. These included widely studied organisms such as the bacterium *Escherichia coli*, the *C. elegans* roundworm, the *Drosophila* fruit fly, and the mouse.

Watson set the official start of the NIH's Human Genome Project as October 1990, with the genome to be completed by 2005 at an expected cost of $3 billion. This proved to be a shrewd and remarkably accurate piece of technological forecasting. The projected budget assumed, correctly, that the cost of sequencing DNA would fall dramatically as techniques improved and efficiencies were gained. The 2005 goal, which seemed highly ambitious at the time, proved to be well attainable.

Another stamp of his stewardship was a program of ethical and legal studies on the consequences of the Human Genome Project. In the past Watson had often taken the lead in saying that the ethical issues raised by new biology should be publicly examined. Many of his fellow biologists, though also concerned about ethical issues, tended to feel that difficult issues were best settled behind closed doors without causing agitation in the press or Congress. Watson was much readier to face the heat of public discussion. He believed that if the future medical benefits of biology were to be accepted, including some of the vexed choices made possible by advances in genetics, biology would have to visibly cleanse itself in the public mind of the stain of eugenics. This was an issue to which he was particularly sensitive because one of his predecessors as director of the Cold Spring Harbor Laboratory, Charles Davenport, had been a leader of the American eugenics movement in the 1930s.

One of Watson's first acts as genome director at the NIH was to announce—without prior consultation with his officials—that 3 percent (later raised to 5 percent) of the NIH genome project's budget would be devoted to ethical, legal, and social issues raised by genome sequencing.[6]

Although the administrative initiative for the Human Genome Project began in the United States, its practical roots lay in Britain, where its foundations were laid by two remarkable scientists, Fred Sanger and Sydney Brenner, both of whom worked in Cambridge for Britain's Medical Research Council. Sanger devised the sequencing method and Brenner the biological context out of which the genome project grew. Born and educated in South Africa, Brenner was for many years a close colleague of Francis Crick at the Medical Research Council's Laboratory of Molecular Biology in Cambridge, England. In the years following the discovery of the structure of DNA, Crick and Brenner figured out, through a remarkably elegant experiment, the way in which the hereditary information is stored in the sequence of bases strung along the backbone of the DNA molecule. The bases form a comma-less code in which each triplet of bases specifies an amino acid, one of the twenty different subunits of which proteins are constructed. Thus GGA codes for glycine, GCA for alanine, TAC for tyrosine and so forth, with the letters A, T, G and C standing for the four different kinds of base in DNA. There is no punctuation between the triplets of bases, so it is critical for the cell's machinery to read the DNA tape in correct phase, meaning always to recognize the first letter of each triplet. If by some error, like the loss of a single base, the machinery should start decoding at the second letter of a triplet, a quite different series of amino acids will be specified and the resulting protein will be unable to perform its assigned task.

Proteins, versatile molecules that perform almost all the practical tasks a living cell must undertake, are linear chains of amino acids folded into complex three-dimensional shapes. The amino acids are linked together by protein-making machines, called ribosomes, that are located outside the nucleus in the periphery of the cell. One of the first puzzles of molecular biology was how the information to build a protein, stored on the DNA in the cell's nucleus, was made accessible to the ribosomes in the periphery. Crick, Brenner and others deduced that the cell must make a transcript of each gene and that the transcript is exported from nucleus to ribosome. These transcripts, which the cell forms from RNA, a close chemical cousin of DNA, are known as messenger RNAs.

These findings, made within a decade or so after the discovery of the structure of DNA in 1953, laid the foundations of molecular biology, the study of the cell in terms of its component molecules. Having solved the central problem of molecular biology with contributions from many other researchers, Crick and Brenner cast around for other issues worthy of their attention and picked the two outstanding problems of biology: how an organism is built from its egg and the organization of the nervous system.

Brenner decided he needed a new experimental animal in which to investigate these problems. The *Drosophila* fruit fly, long the workhorse of geneticists, is just a little speck of an insect yet its brain is too complicated for easy analysis. Brenner picked a minuscule transparent nematode or roundworm as the ideal candidate. *Caenorhabditis elegans,* about a millimeter long, lives in the soil but is no relation of the earthworm. It dines on bacteria and lives about three weeks, a brief allotment of life but one very convenient for the geneticists who study it.

The first major result to emerge from Brenner's *C. elegans* project was a study of its development by his colleague John Sulston. In a tour de force of painstaking microscopy, Sulston mapped the lineage of every cell created in the little animal, from the single cell of the egg to the 959 cells present in the adult worm. Sulston and another colleague, Alan Coulson, then turned to another grand project, that of mapping the worm's genome.

Mapping, for a geneticist, means identifying milestones or markers along the enormous length of a DNA molecule. The DNA in human chromosomes—each chromosome consists of a single DNA molecule, wrapped in the special proteins that package and control it—is about 100 million units in length. The units are known as nucleotides but are often referred to for short as bases—the base is the part of the nucleotide that carries the information—or as "base pairs," because each base is associated with its matching base on the opposite strand. The nucleotide's other two components, substances known as deoxyribose and phosphate, form the backbone of the DNA molecule, with the phosphate unit of one nucleotide linking to the deoxyribose of the next.

The base, also linked to the deoxyribose unit, juts out into the central space between the two DNA strands, making light contact with its counterpart base on the opposing strand. There are four different kinds of bases, the four letters in the DNA alphabet, but two are slightly larger and two are smaller. Because of the precise physical space between the two DNA strands, the presence of a large base on one strand requires that the opposing strand have a small base to pair with it. The two larger bases are called adenine and guanine, the two smaller thymine and cytosine, abbreviated A, G, T, and C. The pairing rule, as discovered by Watson and Crick, is that where one chain has the base A, the other always has T and vice versa; similarly, C and G are always paired.

Just one strand of DNA, called the "sense" strand, carries the genetic message; the other is the "anti-sense" strand because its bases are complementary to those of the sense strand. But genes—lengths of DNA coding for a protein—are found on both chains, so the terms "sense" and "anti-sense" apply just locally, not to an entire DNA chain.

Imagine a book that consists of a single word, about 100 million letters long, with the letters all being A, T, C, or G. That's about the length of a single human chromosome. DNA sequencing machines can decipher the order of the bases in pieces of DNA that are about 500 letters in length. But once the DNA is broken into 500-letter pieces that the machines can analyze, they must be reassembled in order to reconstruct the sequence of the whole chromosome.

The task of reassembly is made much easier by first establishing milestones or markers along the chromosome in the form of recognizable sequences of DNA at known positions. Even a sequence as short as seven letters occurs very seldom within the 3 billion letters of the human genome. Discovering short sequences of DNA at various sites along a chromosome that can serve as markers is a process biologists call mapping.

Working with Robert Waterston of Washington University in Saint Louis, Missouri, one of the many American biologists who had come to work on *C. elegans* in Brenner's lab, Sulston and Coulson succeeded in mapping the roundworm's six chromosomes. The roundworm's whole genome—the length of all its chromosomes added together—is 97 million base pairs of DNA, an enormous length yet barely one thirtieth the size of the human genome. Still, the fact that it could be mapped was the first tangible evidence that it might be possible to sequence anything as vast as an animal genome.

Sulston, Coulson and Waterston first put their worm map on display at a meeting at the Cold Spring Harbor Laboratory in the spring of 1989. The computer print-out, pinned up on wooden slats, stretched right across the back of the room.

Watson, who was then dividing his time between Cold Spring Harbor and his genome office at the NIH, was fascinated with the map. "Jim was delighted," Sulston recalls. Watson called the three of them into his office and worked out a plan to start sequencing the worm's genome. They were to do 3 million bases in three years, with joint funding from Watson's genome budget at the NIH and from the Medical Research Council in Britain.

That would be enough to get started but fell far short of a full commitment to sequence the worm's whole genome. "Jim was properly skeptical—he just was willing to give us enough rope to hang ourselves," Waterston said later. "There was the issue of whether we could accomplish this on the time scale with the methods and strategies we had proposed. Several people told me I was nuts and throwing away my career."[7]

Sulston too was torn between delight at having secured Watson's backing and trepidation at the magnitude of the task they had agreed to perform. He recalls that standing with Waterston on the station platform at Syosset, Long Island, to catch the train back to New York, "I said to Bob, 'The prison door has just closed behind us—I heard it clang.' "[8]

The bet that Watson had laid on the Sulston-Waterston partnership was to prove so successful that the centers run by the two researchers became the backbone of the consortium's effort to sequence the human genome. But before the promise of this transatlantic alliance had emerged, Watson had resigned as director of the NIH's human genome office as the result of a bureaucratic skirmish sparked off by a previously unknown NIH scientist, J. Craig Venter.

Venter had devoted his youth to the serious pursuit of surfing and, in military service, swimming. Sent as a medical corpsman to Da Nang during the war in Vietnam, he was sobered by the experience of so much suffering and chose to become a doctor. He soon decided he could make a better contribution to medicine through research and spent ten years searching for the gene that codes for the receptor in heart muscle cells that detects the hormone adrenaline.

Unlike Watson, who inevitably is somewhat enclosed in the bubble of his own eminence, Venter is affable and approachable. Showing visitors around Celera's plant, he greets janitors and colleagues with the same warmth. Success has not much changed him, except to force him more often from his preferred attire of open-neck shirts and jeans into business suits.

Venter is not a conceptual thinker like most of the biologists in

Watson's circle. He is rather a superb technologist, skilled at seeing what machines can do for him and pushing them to their limits. He has the gift of impatience—he really wants to push problems to their conclusion as fast as technology can take him. He has a knack for seeing the merit in other people's ideas and the self-confidence to make use of them without denying the authors due credit. It was his colleague Hamilton Smith who suggested trying to sequence the genome of the first bacterium, perhaps Venter's greatest single triumph. Smith also suggested the whole genome shotgun method that Venter has used ever since in preference to the mapping strategy followed by the consortium. It was Michael Hunkapiller of Applied Biosystems who invited Venter to race the consortium in sequencing the human genome. The specific approach Venter used to crack the human genome was proposed by two academic researchers, James Weber and Eugene Myers, and rejected by the consortium members to whom it was first offered.

Where other biologists see only risks, Venter sees risks and a way to hedge them, with the result that he has often accomplished feats his competitors judged unlikely to succeed.

Venter's affability to those around him is in marked contrast to the zingers he routinely casts at the consortium's scientists. But sometimes it seems to have been members of the consortium who struck the first blow. A month after Venter announced his intent to sequence the human genome, Maynard Olson of the University of Washington in Seattle assured a congressional committee that Venter's strategy would work to some extent but would "encounter catastrophic problems" when it came to closing the many final gaps in the proposed computer assembly of the genome.[9]

Sulston and Waterston joined in the attack a few months later, explaining to readers of *Science* that Venter's approach "would likely be woefully inadequate."[10]

But the first insult of all, which gave Venter a taste of how roughly the consortium scientists could play when they felt their vital interests were at stake, was hurled by Watson in the public forum of a Senate hearing in July 1991.

Venter was at that time working at the National Institutes of Health and little known beyond the then obscure field of genomics. With his interest in technology, he had formed a close relationship with Applied Biosystems, a California company that had produced the first commercially available DNA sequencing machines by adapting Sanger's manual DNA sequencing method to a machine-readable chemistry. Venter's NIH laboratory had become one of the company's principal testing grounds.

He had put his machines to a typically adventurous use. Most of the genome, at least in animals, is useless DNA that does not code for genes. But it is possible to work with the genes directly by letting the cell, which knows how to read its own genome, find them first. When a gene is activated, the cell makes a transcript of that region of the DNA, and this messenger RNA, after processing, is exported from the cell nucleus to the protein-synthesizing machinery. Venter's approach was to capture the processed messenger RNA transcripts and sequence just the ends of them. These short snippets, though not the full gene, would still be long enough in many cases to identify the gene's role by searching through DNA data banks and comparing the snippets with the DNA of genes of known function. The snippets could also be used to tag the positions of the genes on their parent DNA molecule. Because they come from activated genes, which biologists call "expressed" genes, they are known as expressed sequence tags, or ESTs.

Venter did not invent the EST technique, but he was the first to use it on a large scale, finding several hundred previously unknown genes that were expressed in human brain cells.

The then director of the NIH, Bernadine Healy, decided that the agency must apply for patents on Venter's brain gene sequences, if for no other reason than to prevent other countries from doing so first and to open the debate on whether such partial gene sequences were patentable. Many academic scientists opposed the idea vigorously, saying it would be absurd to allow a patent for a gene whose precise role in the cell was unknown.

Watson also opposed patenting partial sequences, in part because he feared that in a gold rush to patent genes, his consortium would dissolve in squabbling.

At a Senate hearing where Watson was also a witness, Venter made it known that the NIH was seeking patents on the brain genes he had found. Seeking to play down the significance of the patenting issue, Watson declared that it would be sheer lunacy to patent such partial gene sequences and added, quite gratuitously, that the DNA sequencing machines "could be run by monkeys."[11]

The patenting issue created a serious rift between Watson, who had been chosen to head the NIH's genome project by Healy's predecessor, and Healy herself. Healy took various petty bureaucratic actions against Watson, raising questions of a financial conflict of interest that, though judged baseless by higher officials, led Watson to resign in disgust in April 1992.

Venter had no particular stake in the patent issue, in which he was following Healy's directives. A greater source of frustration was that his requests for NIH grants to sequence parts of the human genome had been rejected or deferred by the committee of outside scientists that reviewed genomic proposals. Watson, who could have sidestepped the committee with an internal review, had not done so. Venter, ever impatient, complained in a letter to Watson, "I am concerned that the bureaucracy that is a necessary part of the grant review process cannot keep pace with the rapid developments in the area."[12]

Venter was about to be liberated from the delays of bureaucracy and the obstruction of scientists in the genome establishment. Wallace H. Steinberg, a venture capitalist and chairman of the board of HealthCare Investment, was intrigued by the possibilities of Venter's EST method and proposed to set up a company to pursue it. Besides his commercial motive, Steinberg also cast his interest in nationalistic terms: "I suddenly said to myself, 'My God—if this thing doesn't get done in a substantive way in the United States, that is the end of biotechnology in the U.S.' "[13]

But Venter, still regarding himself as an academic researcher, insisted on a nonprofit institute. So Steinberg set up a curious partnership: Venter was to run the nonprofit Institute for Genomic Research, which would be supported with a $70 million grant from a new company, Human Genome Sciences, that would have commercial rights to Venter's discoveries. To run Human Genome Sciences, Steinberg chose a Harvard Medical School AIDS researcher, William Haseltine.

In July 1992, three months after Watson's resignation, Venter also left the NIH, taking almost his entire staff of thirty people with him, and from that point the two men followed separate but intermittently clashing paths. At the Institute for Genomic Research, for which Venter chose the aggressive acronym TIGR, he at first continued his EST approach, eventually capturing ESTs from most human genes. This achievement brought him almost as much grief as gratification. Academic scientists were allowed access to TIGR's EST database but chafed at having to show their manuscripts to Human Genome Sciences before publication in case it wished to patent anything. Backers of the consortium were also unhappy with the EST approach, despite the quick route it offered to novel genes. Their concern was that Congress might become unwilling to pay for sequencing the whole human genome if Venter were to claim he had fished out all the interesting parts, namely the genes.

In 1993 Venter's interests took a different turn. He had become friends with Hamilton O. Smith, a biologist at Johns Hopkins University who had won the Nobel Prize in 1978 for discovering restriction enzymes, a wonderfully cunning defense mechanism produced by bacteria against the viruses that prey on them. (Restriction enzymes cut DNA at specific short sequences of bases; the host bacterium's DNA, however, has evolved to lack the specific sequence recognized by its restriction enzyme, so the enzyme chops up all DNA in sight except its own. Restriction enzymes are invaluable tools for biologists because they can be extracted from bacteria and used as chemical scissors for snipping DNA at chosen sites.)

Smith suggested that Venter try to sequence the genome of a bacterium. At that time the longest piece of DNA that had been sequenced was a 100,000 base pair stretch from *C. elegans,* a triumph that Sulston and Waterston had published the year before. Most bacterial genomes were 1 million base pairs and up. Under Watson the NIH's genome office had chosen to sequence a bacterium called *Escherichia coli* because it was a laboratory favorite. But its genome was enormous—4.6 million base pairs—and the University of Wisconsin team that had been assigned the task was making slow progress.

Smith proposed that they tackle his own favorite bacterium, one called *Haemophilus influenzae* but, despite its name, no relation to the flu virus. Its genome was a much more manageable 1.8 million bases in length. Smith also proposed a different sequencing strategy. Instead of first mapping the genome by finding marker sequences all the way along it, as Sulston and Waterston were doing for *C. elegans*, Smith's idea was to skip the time-consuming mapping stage altogether by just breaking the genomic DNA into random pieces, doing it many times over so as to get a different pattern of breaks each time, and ensuring there were enough overlaps between the pieces for a computer to be able to piece everything together again.

The method became known as a whole genome shotgun because the entire genome is first blasted into small pieces, each of which is then amplified, or cloned, by being inserted into a bacterium. After each bacterium has multiplied, the inserted DNA is retrieved in a sufficient number of copies to allow further analysis by the DNA sequencing machines.

Venter announced at a meeting in May 1995 that he had successfully sequenced the entire *Haemophilus* genome. The news had a broad impact in the scientific community, spreading his reputation beyond the small world of genome sequencers. As the first bacterial genome to be sequenced, the *Haemophilus* sequence gave biologists their first glimpse into the basic genetic mechanisms required by a free-living organism, and thereby into the nature of the living cell. (The genomes of many viruses had already been sequenced but viruses are not free living, independent entities; their genetic material ranges only from 5,000 to 200,000 or so units in length and contains a mere handful of genes, which constitute simple though cunning recipes for hijacking a cell's machinery.)

Venter's declaration was particularly dramatic because it came out of the blue. He had not gone around talking about interim results at scientific meetings, as is the usual academic practice. He waited until he could unveil a fait accompli at the main annual conference of bacteria experts, a meeting of the American Society for Microbiology. With a flair for the throwaway line, he also mentioned at the meeting that he had completed another genome, that of the bacterium *Mycoplasma genitalium,* of which more details would be furnished later.

Venter has developed the practice of announcing some lofty goal, waiting for genome sequencers in the consortium to predict its certain failure, and hiding his hand until he can announce, out of the blue, that the goal has been achieved. In the case of the *Haemophilus* genome, he applied for a grant to the NIH. The work went so quickly that he had already completed about 90 percent of the genome when he received the NIH's reply: his request was refused because the review panel, a body that included several of the consortium's genome experts, had judged that his approach was so unlikely to succeed that it was not worth supporting.[14]

The victory was doubtless doubly sweet because the NIH's own bid to sequence the first bacterial genome, that of the *Escherichia coli* bacterium, which lives in the human gut, was way behind schedule. By choosing a more appropriate target, and a riskier but more promising method, Venter had snatched a great prize from the bureaucracy that had thwarted him.

Other scientists were quick to acknowledge his achievement. "This is really an incredible moment in history" was the generous verdict of Frederick R. Blattner of the University of Wisconsin, the leader of the NIH's *Escherichia coli* project. "It demonstrates the ability to take the whole sequence of an organism and work down from that to its genes, which is what geneticists have been dreaming of for a long time."

Venter, eager perhaps for Watson's approval, pointed out to reporters an essay in which Watson had said what a great thing it would be to crack a microbe's genome. "Just getting the complete description of a bacterium—say, the five million bases of *E. coli*—would make an extraordinary moment in history," Watson had written just three years previously.[15] Watson did praise Venter's achievement, calling the sequencing of the *Haemophilus* genome "a great moment in science" and noting that "With a thousand genes identified, we are beginning to see what a cell is."[16] But even this success did not draw Venter into the magic circle of Watson's approved biologists.

The article reporting the *Haemophilus* genome sequence has become one of the most frequently cited papers in the scientific literature.[17]

With the success and attention that followed, Venter found it easier to raise funds for his institute. He and Haseltine, the chief executive of Human Genome Sciences, did not enjoy an entirely cordial relationship. Haseltine, formerly a geneticist at the Harvard Medical School and an expert on the AIDS virus, is a strong personality, suave and articulate, with an interest in power and influence. His wife, Gale Hayman, built a cosmetics company known for the best selling fragrance Giorgio Beverly Hills. He is a trustee of the Brookings Institution, a Washington political think tank. Though both men have become multimillionaires, Venter retains the common touch while Haseltine more openly enjoys the trappings of wealth, letting himself be photographed in his limousine.

Haseltine considers himself as by far the keener of the two to apply genomic knowledge to health. He is directly developing drugs and already has products in clinical trials whereas Venter, he alleges, is a restless spirit, always moving from one thing to another and most deeply interested in comparative genomics, the relationships between the genomes of different species. Venter, for his part, makes the same point about Haseltine's sweeping claims of success as do many academic biologists: that no one can tell what Haseltine has achieved because his major claims have not been published.

After the death of their patron, Wallace Steinberg, in July 1995, there was no one to compel harmony between the two. Haseltine's purpose was to use Venter's EST approach for discovering gene products that would make useful drugs. He intended to do this ahead of the slower moving established pharmaceutical companies and to build his company into a powerhouse of genome-based medicine.

Venter, still looking more toward the academic world, preferred to continue exploring the world of genomes. He also resented the commercial link with Human Genome Sciences, which made his institute suspect in some academic quarters. In 1997 TIGR and Human Genome Sciences severed all ties. To be liberated from Haseltine, Venter was willing to renounce no less than $38 million. That was the sum that Haseltine was still obliged to pay as part of TIGR's endowment and that Venter forfeited in the divorce.[18]

To appreciate the genesis of Venter's next coup, it is helpful to trace another thread in genome sequencing, that of the machines that work out the order of bases in a DNA molecule. Like the roundworm that paved the way for the Human Genome Project, the sequencing machines had their origin in the golden circle of biologists who worked at the Medical Research Council's Laboratory of Molecular Biology in Cambridge, England.

The first genome ever sequenced was that of a virus called ΦX174. The order of its 5,375 (later revised to 5,386) nucleotides was determined in 1977, after a thirteen-year effort, by Fred Sanger of the Medical Research Council's Cambridge laboratory. Sanger is less well known than Watson and Crick but his stature in biology is almost as large because he worked out chemical methods for determining the order of the subunits in both proteins and DNA, receiving a Nobel Prize for each achievement. He was the first to show that proteins are made of strings of amino acids linked head to tail. Turning to DNA, the other main kind of macromolecule in living cells, he developed a wonderfully ingenious method for sequencing the giant molecule, using the ΦX174 virus as his experimental target.

His method, an automated form of which is used by today's DNA sequencing machines, depends on having the target piece of DNA copied many times over by a DNA copying enzyme. The enzyme is supplied with the four nucleotides of which DNA strands are made, but the stock of one of the nucleotides, say adenine, contains a pinch of mock adenine. The mock chemical is like adenine in every way save that it cannot be linked to the next nucleotide in the growing DNA strand.

If the DNA copying enzyme were given 100 percent mock adenine, it would start making a new DNA strand and terminate the strand as soon as it came to the first A. But with just a pinch of mock adenine in its nucleotide stock, the enzyme will make strands of varying lengths, stopping each time at whatever point it happens to pluck a mock adenine from its stock instead of the real thing. In fact, in the right conditions, the enzyme will make a full set of all possible strands, each member of which was stopped at one of the positions where there is an adenine in the DNA strand being copied. This set of incomplete DNA strands contains vital information: the position of every adenine in the test DNA.

That information is easily visualized by forcing the DNA strands to travel through a slab of gelatin-like material. The strands come to rest at different positions depending on their length, with the shortest strands traveling farthest. And because the mock adenine nucleotides are synthesized from radioactive atoms, each strand signals its position by leaving a black bar on a photographic plate placed under the gel.

In parallel tracks on the gel, three other sets of incomplete DNA strands are separated, corresponding to the guanine, thymine, and cytosine nucleotides. All that the researcher now has to do is to read the black bars in sequential order in order to infer the sequence of nucleotides in the test DNA. If the fastest and shortest chain was in the lane for guanine, the second shortest in the cytosine lane, the third also in cytosine and the fourth in adenine, then the sequence of the copied DNA molecule begins G-C-C-A-.

The mock nucleotides are made with a chemical called dideoxyribose in place of deoxyribose; hence Sanger's method is called dideoxy sequencing. The invention was a heroic achievement. But it was not powerful enough to sequence anything much larger than the genomes of viruses and mitochondria, the energy-producing subunits of animal cells. Even the genome of a simple bacterium was far too large for Sanger's method. As for tackling the human genome with this labor-intensive manual technique, the task would have taken a team of 1,500 scientists one hundred years.[19]

Large-scale DNA sequencing marked time for more than a decade, the interval required for technology to advance and people to work out methods of automating Sanger's method. The pioneers of this effort were a team of biologists and engineers working at the California Institute of Technology under Leroy Hood and at Applied Biosystems, a company founded in 1981 to exploit the Hood team's inventions.

One critical step was to replace the radioactive atoms with which Sanger labeled his dideoxy nucleotides in favor of fluorescent dyes of four different colors, one for each of the bases. Though the dyes were typically present in minute amounts, they could be activated by a laser beam and made visible to a machine's optical scanner. Members of the team that developed the first Applied Biosystems DNA sequencing machine included Hood, Lloyd M. Smith, and the brothers Michael and Tim Hunkapiller.[20]

Independently, a fluorescent dye sequencer was developed by the Swedish company LKB-Pharmacia.

One mark of the Hood team's foresight is that they conceived and designed the DNA sequencing machine before biologists realized how much they needed it. "There was a lot of skepticism early on as to whether automation was necessary," Michael Hunkapiller says. "DNA sequencing was a fairly successful manual process. People didn't appreciate the value of automation nearly as much as for proteins," for which the Hood team had also developed a sequencer.[21]

As an avid technology user, Venter had tested out the first Applied Biosystems DNA sequencers while still at the NIH. The first generation of machines separated DNA, as Sanger had done, on a slab of gelatin-like material made of a toxic material called polyacrylamide. The slab gel machines were reliable workhorses for genome biologists throughout the early and mid-1990s. But they could cause accuracy problems when separate lanes in the slab gels got mixed up. And preparing the gels themselves was a labor-intensive task that required large numbers of expensive technicians.

Applied Biosystems started working on a new model of sequencer that separated DNA by running the chains through very fine tubes instead of through a slab of acrylamide gel. The capillary tubes avoided the problem of crossed lanes. The machines were also cheaper to run because of economies in labor and materials.

In the fall of 1997, as the capillary sequencers neared production, Michael Hunkapiller, now the president of Applied Biosystems, realized they had the potential to far outstrip the capability of their slab gel predecessors. The machines required no time-consuming pouring of gels and could run for twenty-four hours with only fifteen minutes' supervision.

The public consortium had not deviated from the goal and timetable laid down by Watson in 1990. It expected to complete the genome by 2005, with most of the sequencing to be done in the final years. As of early 1998, only 3 percent of the genome had been sequenced. Hunkapiller believed that with a dedicated suite of the new capillary sequencing machines, later to be called Prism 3700s, it should be possible to sequence the human genome much sooner, if done by the right person.

Having worked with Venter for a decade and noted his management skills in setting up TIGR, Hunkapiller recognized the ideal person to implement his ambitious scheme. In January 1998, he invited Venter and Mark Adams, a close colleague of Venter, to Applied Biosystems' drab office buildings in Foster City, at the back of San Francisco airport. "We approached Craig in January along with Adams, and said, 'We'd really like you to come out and discuss a grand vision of what you might do it,' " Hunkapiller recalls.

"We said, 'You could sequence the whole genome.' Craig said, 'You gotta be crazy.' We spent a few days working through the math and came away thinking 'Maybe it is doable.' They went away and redid the calculations, and so did we."[22]

Hunkapiller's laconic engineer's style glosses over the enormous risks both men were about to take. Hunkapiller's was that his customers in the consortium would object to their supplier going into competition with them and flock to the rival capillary sequencing machine then being developed by Pharmacia. Applied Biosystems' parent company, Perkin-Elmer, would invest an initial $75 million in a new division to sequence the genome. But easing the risk was the fact that in initiating a rival genome effort Hunkapiller was doubling the potential size of his market. Moreover the Prism 3700s were to sell for $300,000 apiece, twice the price of their predecessor, and doubtless carried a reasonable profit margin.

For Venter, the human genome was 1,500 times as large as *Haemophilus*'s, and there was no guarantee that its special problems, including numerous repetitive sequences of DNA, would be soluble at all. To accept Hunkapiller's offer he would have to leave the security of his own nonprofit institute and leap into a race with the consortium, a competitor that enjoyed a decade's start, the backing of the U.S. government and the world's richest medical charity, and the expertise of the world's leading biologists.

But considering the heaven-sent opportunity to snatch the human genome from the consortium's grasp, how could he resist? Venter had long felt that the human genome could be sequenced faster than the consortium was doing it. He believed that the same whole genome shotgun approach with which he and Smith had successfully cracked the *Haemophilus* genome would also work on the human genome. The shotgun approach would sidestep the arduous and error-prone mapping process used by the consortium. It would require sequencing the genome many times over, so as to be sure of having enough overlaps between DNA fragments to piece the DNA back together again. A major problem would be the repetitive sequences that riddle the human genome, since computer assembly programs are easily confused by a stretch of DNA in which the same sequence is repeated many times over.

Without yet having an exact solution to this problem, Venter, Adams and Hunkapiller figured they would need to sequence the human genome ten times over. With the human genome being about 3 billion base pairs in length, and given that the sequencing machines could analyze fragments of DNA about 500 base pairs long, they would need to generate 60 million fragments. They calculated that, with two hundred or so of the new machines, they could sequence the human genome within three years, in other words before the end of 2000.

A new division of Perkin-Elmer would be set up to sequence the genome. It would later be named Celera, after the Latin word *celer* meaning quick. But how would it make money? The business plan developed by Hunkapiller, Venter and Tony White, the Perkin-Elmer chief executive who had decided to plunge the scientific instrument maker into genomics, was for the division to become an information company. It would sell subscriptions to its database of the human genome and later many other genomes, and would provide state of the art programs for analyzing the data. Although it would seek patents on a few select genes, it would not attempt to lock up broad swathes of the genome in the style of other genomics companies such as Incyte and Human Genome Sciences.

The plan was tightly held until leaked to *The New York Times* on May 10. Venter gave the consortium's leadership two days' warning of his plans by informing Francis Collins, Watson's successor as director of the NIH's Genome Office, and Harold Varmus, director of the NIH.

The two men's first reaction was that they would accommodate the consortium's program to Venter's initiative, letting him bear the brunt of sequencing the human genome and moving the academic biologists supported by the NIH on to the next and equally important step of interpreting the genome. "Our plate is overly full, and the resources to do this have appeared to be on the edge of what is needed," Collins said in initial reaction to Venter's plan. "Our plan is to integrate the government program with Craig's." The NIH could "pursue other model organisms like the mouse," he said.

But would Congress accept the idea of yielding the human genome to Venter and settling for the lesser glory of sequencing the mouse genome? "We have told Congress that the genome project is many more things than getting the human genome. But it will probably require some explanation and education of why it is that this is an opportunity," Collins said.[23]

Varmus too welcomed the implicit division of labor made possible by the news that Venter would take over the chore of sequencing the human genome while the NIH's grantees moved on to the more interesting task of interpreting it. "It seems to me that this is an example of how a technical advance in the private sector can move the whole enterprise forward faster," he said. So what would happen to the consortium's effort to sequence the human genome? "There are many other genomes to sequence. From my point of view it means we move forward faster in sequencing related organisms like the mouse, the zebra fish, the rat, and comparative primates."[24]

As Venter knew in timing his announcement for maximum impact, the consortium's genome sequencers were holding their annual meeting that very week at Watson's Cold Spring Harbor sanctuary. They were far from sharing the welcome given to Venter's initiative by Collins and Varmus. The human genome, in their view, was a great scientific prize, not a chore. They did not rush to embrace the idea of "integrating" their efforts with Venter's, let alone subordinating them to his leadership. They suggested all kinds of reasons why Venter's effort was likely to fail. Because of the shotgun strategy, the quality of Venter's sequence would be "very significantly compromised," Waterston said, with the final product being similar to "an encyclopedia ripped to shreds and scattered on the floor."

Pressure began to build on Collins to hold steady on course and to eat his words about integration with the enemy. John Sulston flew in from Cambridge, bringing his program officer from the Wellcome Trust, Michael J. Morgan. Referring like his colleague Bob Waterston to the gaps he thought a shotgun strategy would produce, Sulston said of Venter's plan that "I really don't see this as being any great advance whatever. We are going to provide the complete archival product and not an intermediate, transitory version of it."

The Wellcome Trust shared his view that it would be unthinkable for the consortium to abandon the human genome to Venter's company. "To leave this to a private company, which has to make money, seems to me completely and utterly stupid," Morgan said.

Asked if the trust were prepared to finance sequencing the entire genome if the NIH should pull out, Morgan replied, "If we had to and if we wanted to, we could do it." The trust's assets were at that moment worth $19 billion, he noted. The long latent trust, founded by the American pharmacist Henry S. Wellcome when he moved to London in 1880, had suddenly become rich by overriding a stipulation in Wellcome's will and selling its major asset, the Burroughs Wellcome pharmaceutical company, which had struck gold with its AIDS drug, AZT. There was no little irony in the fact that Burroughs Wellcome's bonanza with AZT, a drug for which it initially charged $10,000 a year, was the springboard of the Wellcome Trust's sudden fortune and the means whereby Morgan could both support the public consortium and denounce the greed of private companies.

The British visitors were intent on doing everything possible to stiffen the spine of their vacillating ally. Collins was hard put to maintain his initial attitude of cooperation with Venter when Waterston and Sulston, the backbone of the consortium, were adamantly opposed to any such idea.

The coup de grâce was delivered by Watson. In his customary understated way, he compared Venter's assault on the genome with Hitler's annexations and demanded to know of Collins whether he intended to play Chamberlain's role or Churchill's.[25]

While the consortium's scientists were reeling from the news of Venter's putsch, he arrived at Cold Spring Harbor to drop a second bombshell on his stunned audience.

Just as the consortium was sequencing the roundworm's genome as a pilot project for its human genome strategy, even Venter considered it would be prudent to first try out the whole genome shotgun approach on some animal genome less demanding than the human. A desirable subject would be the *Drosophila* fruit fly, the laboratory workhorse of geneticists for ninety years.

Because of biologists' intense interest in seeing the fruit fly's genome sequence, Watson had included *Drosophila* among the model animal genomes being sequenced as part of the Human Genome Project, and a *Drosophila* project had started under Gerald Rubin, a leading fruit fly biologist at the University of California, Berkeley. Venter announced at the Cold Spring Harbor conference that he too was going to sequence the fruit fly genome. He then invited Rubin, who was in the audience, to collaborate with him.

Despite the casual nature of the public invitation, Rubin accepted. Being a geneticist, not a full-time genome sequencer, he just wanted to get the fruit fly's genome done, and if Venter wished to bear the brunt of the sequencing, that was fine with Rubin.

After the May 1998 Cold Spring Harbor meeting, it was clear that both human genome efforts would continue and that a furious race lay ahead. Collins, a medical geneticist by background, may have preferred cooperation—he kept assuring everyone that there was no race—but the requirement for holding the consortium together was to promise to lead it to a victory of some kind.

In December 1998 the consortium scored a major scientific triumph when the Sulston-Waterston team finished the genome of the roundworm, the first animal genome to be sequenced. The little creature turned out to have far more genes than had been expected, revealing an unimagined level of complexity to biologists taking their first look at what it takes to develop, operate and maintain an animal. But there were many small, intractable gaps in the sequence. Though none was important, the gaps confirmed that completeness would be harder to attain in sequencing animal genomes than those of bacteria.

On the human genome front, Sulston and Waterston ramped up production of human DNA sequence as the worm work phased down. But it was clear that the original 2005 target date set by Watson needed to be brought forward, just in case Venter's version of the genome was not as woefully inadequate as they predicted. If Venter were to complete the genome before 2001 and do a good job of it, the consortium could have little to show for its efforts.

Collins then listened to an old idea that now resurfaced in a different form, that of going after the genes first. The DNA that codes for genes takes up only 3 percent of the genome, and the genes are not evenly distributed but bunched together in gene-rich regions with semi-desert stretches of DNA in between. So it would be easy to target the genes first, and since these were of interest to medical geneticists, the consortium would be able to satisfy a major constituency much sooner than otherwise.

The genome sequencers had long resisted this obvious idea for fear that if the genes were fished out first Congress might not understand why it was paying to sequence the other 97 percent of the genome. (Biologists would have said that many answers about human evolution and possible surprises lie there.) But the urgency of Celera's challenge dictated a new strategy. In September 1998 Collins decreed that as an interim goal, gene-rich regions containing one half of the genes, and amounting to one third of the genome, would be completed by the end of 2000. The consortium would also move up its goal for completion by two years, from 2005 to 2003.[26]

Six months later, with the consortium's sequencing efforts making good progress, Collins felt able to declare a new interim deadline. The consortium would produce a first draft of the human genome, covering 90 percent of the DNA, by the spring of 2000. This was adroit timing because, if the consortium could achieve it, production of the draft human genome sequence would steal the thunder from whatever Venter might achieve by the end of the year. For someone not engaged in a race, Collins was not without guile in seizing the lead.

At the same time he shed much of Watson's original scheme of spreading the genome money around many centers and focused a round of extra financing on the most productive, cutting loose the less efficient. This was a necessary step if the consortium were to have any hope of matching Venter's economies of scale. Genome sequencing is not an ordinary academic occupation; it requires special management skills and the conversion of one's laboratory into an industrial-scale plant. By March 1998, Waterston's genome sequencing center employed two hundred people working in shifts and operated nineteen hours a day.

Collins announced in March 1999 that he would focus the NIH's part of the consortium's efforts on three centers: Waterston's at Washington University, Saint Louis; Eric Lander's at the Whitehead Institute in Cambridge, Massachusetts; and Richard Gibb's at the Baylor College of Medicine in Houston. The other principal members of the consortium were the Department of Energy's Joint Genome Institute in Walnut Creek, California, and the Sanger Centre in England. The consortium's total cost for preparing the draft human genome sequence would be $280 million, Collins estimated, stressing that the $3 billion usually cited as the Human Genome Project's full fifteen-year cost included many other items such as infrastructure and sequencing the model organism genomes. The $280 million figure put the consortium in the same ballpark as Venter, who had declared he would sequence the human genome for only $200 million.[27]

With Collins's new approach, the consortium seemed to have gotten itself in good shape for the race to the draft sequence. Venter professed not to see it that way. "Both Morgan and Collins are putting good money after bad," he said in an expansive mood later that year, as the first fifty Prism 3700 DNA sequencers and an enormous new supercomputer—the second most powerful in the country, according to Compaq, its maker—were being installed in Celera's new plant in Rockville, Maryland.[28] But Collins, addressing the consortium's troops a week later at their annual conference in Cold Spring Harbor, was able to report that 10 percent of the genome had been completed and that the consortium was on track to finish its first draft in a year's time as promised.

With Watson sitting by, Collins described the consortium's enterprise as "the most important organized scientific effort that humankind has ever attempted. It dwarfs going to the moon." The scientific meeting took on the overtones of a pep rally as a projection of Venter's image drew laughter from the crowd. Watson deemed the consortium's progress to be "extremely satisfying."[29]

By December 1999, the consortium was able to report that it had completed the sequence of the first human chromosome, number 22, one of the smallest of the 23 pairs in each human cell. "A new era has dawned—we have fulfilled the dreams of Mendel, Morgan, Watson and Crick," proclaimed Bruce Rose of the University of Oklahoma, one of the contributors to the chromosome 22 structure. There were a few small gaps, to be filled in later, and a certain kind of repetitive DNA, known as centromeric DNA, proved unsequenceable by present methods. The centromeric DNA, located toward the center of each chromosome, helps the two parts of a duplicated chromosome separate when the cell divides; its role seems to be largely structural, and it contains few or no genes.

But as the consortium racked up its first signs of solid progress, a complex political intrigue was playing out behind the scenes. Celera's supercomputer was spitting out its first assemblies of the *Drosophila* genome under the guidance of Eugene Myers, the University of Arizona mathematician whose proposed computational method for sequencing the human genome had been rejected by the consortium. When Myers had heard of Venter's proposal to sequence the genome, he had called up and insisted he be on the team. Myers, along with Granger Sutton, soon assumed a critical position as Celera's chief software architect.

Myers's success in getting Celera's supercomputer to puzzle out the *Drosophila* genome changed the power balance between the two sides. The major problem in sequencing animal genomes is the repetitive DNA, regions where the same, or roughly the same, sequence of nucleotides is repeated over and over. The repeats may be just two nucleotides or several thousand in length. Much like identical pieces in a jigsaw, the repeats create ambiguity that frustrates the search for a unique solution.

When the computer assembly program tries to piece together a genome from 500-base-long fragments—the longest sequence of nucleotides that the DNA sequencing machines can analyze—it looks for unambiguous overlaps between them. The program soon joins the fragments into long collections of overlapping or contiguous pieces, known for that reason as contigs. But then the assembly program gets stuck, because between the contigs lie the more or less identical pieces of some repetitive sequence. Without further information, the assembly program cannot decide how the various contigs are arranged in relationship to one another or how many repeats lie between the end of one contig and the beginning of the next. This is the principal reason why the critics of Venter's whole genome shotgun strategy predicted its catastrophic failure. The bacterial genomes on which Venter's shotgun strategy had worked so well do not possess the confounding repetitive DNA.

The consortium's way around this problem was to break the human genome down into a large collection of fragments known as bacterial artificial chromosomes, or BACs, figure out the sequence of the DNA units within each BAC, and then reassemble the whole genome from an overlapping series of BACs.

The BACs are themselves rather large chunks of DNA, about 150,000 bases or so in length. To generate enough of each BAC fragment for the sequencing machines to analyze, the fragments are inserted into bacteria and each bacterium is allowed to divide many times. After this growth process, each colony of bacteria contains clones—identical copies—of its founder bacterium and, more important, of the inserted fragment of DNA, which is then extracted. These cloned fragments of human DNA are known as bacterial artificial chromosomes (BACs) because the bacterium is tricked into duplicating the fragment along with its own chromosome.

DNA sequencing machines determine the order of a BAC's nucleotides in 500-base-long pieces, and a computer program assembles the pieces into a full length BAC. Repetitive DNA sequences are also a problem for the BAC method but are confined within the 150,000-base section and hence are much more manageable.

Where on its parent chromosome does each BAC belong? The computer simply looks for known markers—short sections of DNA of known sequence—along the BAC. If two BACs share the same pattern of markers, they obviously overlap. And the markers in each individual BAC will show its position on its parent chromosome.

The mapped BAC strategy is in principle safe and sound. In practice it is difficult and time-consuming. But how could a genome be assembled without the positional information provided by the BAC map?

The Venter-Myers approach rested in part on the ability of their colleague Hamilton Smith to prepare cloned libraries of DNA fragments. One of Smith's skills was to prepare fragments of precise lengths, such as 2,000, 10,000, and 50,000 bases. The 10,000-base pieces of DNA are longer than most stretches of repetitive DNA. Myers could therefore use them to link his contigs together. The sequencing machines analyzed just the ends of the 10,000-base fragments. The assembly program would look for pairs of contigs each of which contained a different end of the same 10,000-base span. Bridging a pair of contigs in this way defined the length of the gap between them and enabled the program to figure out how many repeats the gap contained. The linked contigs grew into structures that Myers called scaffolds. When the Celera supercomputer started spewing out immense scaffolds, some as long as a whole *Drosophila* chromosome arm, he and Venter knew for the first time that the whole genome shotgun strategy would actually work.

By October 1999, Myers was able to say that he possessed 97 percent of the fruit fly's genome, including almost all of its many known genes, in correctly assembled contigs. "We are absolutely not getting lost or fooled by the repeats—we are navigating right around those babies," he said, declaring his critics refuted.[30]

As news of this achievement spread, the logic of Collins's initial instinct for cooperation with Celera became more evident to some members of the consortium, particularly Eric Lander. Because the consortium and Celera had followed different strategies, their data were independent and could very fruitfully be combined. By exchanging the raw electrophoretic data from the sequencing machines—the evidence on which each nucleotide was determined—the two sides could help each other figure out tricky regions of the DNA that did not amplify well. And what if Venter was as good as his word and actually beat the consortium hands down? Two senior biologists on Celera's scientific board, microbiologist Norton Zinder of Rockefeller University and Richard Roberts, a Nobel Prize winner at New England Biolabs, believed from their inside knowledge that Celera was way ahead but that a public defeat for the consortium would harm the NIH and Congress's appetite for giving big money to biomedical research. They too started urging collaboration between the two teams.

As a result of an approach from Lander to Venter, a meeting between the principals was held on December 29 to discuss a possible merging of the two sides' genome data and a joint publication of the results. The consortium was represented by Collins, NIH director Varmus, Waterston, and a representative from the British side. The Celera team included Venter and Tony White, the chief executive of Celera's parent, by then renamed the PE Corporation.

But the meeting quickly bogged down over the issue of who should control the merged data. The consortium demanded that its access rules should prevail—the data should be instantly free to everyone. But Celera, whose business plan was to create a genomic data bank based on the human genome, insisted on some commercial protection for its data, particularly rules to bar rival genomic companies from downloading Celera's data for free and repackaging it for sale.

The consortium submitted a list of "shared principles," drafted by Lander, which included the lofty thought that "humankind will be better served if we can find a viable way to join forces," the wistful hope that "the current antagonism and excessive competition should be replaced with a more collaborative spirit," and the admonition that agreement would require "avoiding disparaging the other party, sowing discord or undermining the collaborative spirit."

But both Lander and the collaborative spirit were excluded from the meeting, which ended without agreement. Its only result was to increase the antagonism between the two sides. Humankind was put on hold.

Venter, meanwhile, was preparing to take another of his breathtaking gambles, which was to reduce Celera's planned coverage of the human genome from tenfold to fivefold. The first step in sequencing a genome is to break an organism's whole DNA into fragments and insert the fragments into bacteria for cloning and amplification. Because of the random nature of the cloning process, the library of cloned fragments needs to be many times the length of the genome in order to be sure that every part of the genome is included. A library whose DNA is one genome in length, when each of the different fragments is added together, has only a 63 percent chance of including any specific piece of DNA from the genome. To be more than 99.99 percent certain of getting every single piece of DNA into the library, it's desirable to shoot for tenfold coverage.[31] In other words, for a genome of 3 billion base pairs, like the human genome, 30 billion bases need to be sequenced.

Venter and Hunkapiller's original intention in May 1998 had been to shoot for tenfold coverage. But Collins's plan to announce a first draft of the genome in spring of 2000, preempting Celera's plans for completing the genome by the end of the year, had caught Venter's attention. He did not like to be upstaged. There was doubtless financial pressure on him too. Though Celera was well funded, it was burning money fast; to run his supercomputer and three hundred Prism 3700s, Venter's electricity bills alone came to $100,000 a month. The sooner the human genome database could be offered to paying subscribers, the better. Moreover it was becoming clear that Celera needed to sequence the mouse genome as well and offer that in a package with the human genome.

The mouse genome was emerging as an invaluable interpretive tool, because by laying the two genomes side by side one could tell at once where the genes were probably located. The reason has to do with evolutionary change. Mice and men, both being mammals, have a very similar set of genes and genomes of about the same size. In the course of the 100 million years since the forebears of mice and people split apart from their common mammalian ancestor, the DNA bases in their genes have been subject to continual mutation through the usual forces of copying errors, radiation, and spontaneous chemical change. The various elements of the genome react differently to this mutation pressure. The DNA sequence of the genes cannot change very much. If it does, it will in most cases produce a dysfunctional protein and the animal that contains it will die without progeny. But the non-coding regions of the genome, the repetitive DNA sequences that perform no useful role, are free to mutate without limit.

In the course of time, therefore, the non-coding regions of mouse and human DNA have grown progressively different from each other, whereas the protein-coding regions of DNA have been forced to remain quite similar. The result is pretty much that the parts of human and mouse DNA that resemble each other are either genes or the control regions of DNA that govern the genes' activity, and parts that don't resemble each other are DNA regions of little practical interest. Hence the desirability of sequencing the mouse genome to help interpret the human genome.

In January 2000, just after the aborted peace meeting with the consortium, Venter announced that he had sequenced 5.3 billion bases of human DNA and was ready to proceed to the assembly phase.[32] With less than twofold coverage, he was likely to be missing 15 percent of the genome altogether. What could he be thinking of?

As usual, Venter had carefully gauged his risks, but the consequences of this calculation sent the consortium into a froth of rage. The consortium posted its new human sequence data every night on the DNA data banks for free use by everyone. Venter assumed that that included even him. By combining the consortium's data with his own, he would by April 2000 have about 4.6-fold coverage, which would give a 99 percent chance of including every piece of DNA in the genome. That meant he could pre-empt Collins's rough draft of the human genome with an essentially complete genome from Celera and then switch all his sequencing machines over to the mouse genome, the quicker to offer both mouse and human genomes to his subscribers.

Consortium scientists were aghast. They wanted to publish their own data in a scientific journal, not have Venter do so for them. For Celera to sequence the human genome first would be bad enough; to have Venter scoop them with their own data would be insufferable.

In a letter of February 28, 2000, Collins, Varmus and Waterston wrote to Venter. "Publication of other groups' primary data without their consent is considered to be a breach of scientific ethics. In addition, it is counter to accepted scientific practice for authors to present results without having examined the primary data."

Though the letter was signed by two senior NIH officials, the heavy charge was of doubtful merit. In terms of scientific credit, submitting data to GenBank was generally considered a publication, so the consortium had already published its data. Moreover it was hardly consistent to make a virtue of posting genomic data for everyone's free use and then to assert that "everyone" meant everyone but Venter. The fact that Celera could see and use the consortium's data but not vice versa was indeed unfair in terms of the race. But the unfair advantage was a direct consequence of the consortium's adamant policy on immediate data release. And this policy, though admirable, was highly unusual in that scientists rarely make their raw data instantly available, preferring instead to publish it along with their own interpretation.

The Collins-Varmus-Waterston letter sought to cast blame on Celera for the failure of the December 29 meeting, complaining that Collins had been unable to reach Venter by phone for a week, and warned that unless Venter replied within a week to deny their charges, the proposed merger of data would be off.

Waiting until the ultimatum had expired, Venter replied on March 7 saying, in effect, that he had a perfectly amicable collaboration going with Gerald Rubin, a consortium-supported scientist, on the fruit fly genome, and that if the human genome side of the consortium couldn't be as reasonable as Gerry, that was their problem.

Venter's point was true but misleading. Rubin could collaborate easily with Venter because he was neither a genome scientist nor a personal rival. And the fruit fly genome, though fascinating to biologists, was not nearly as valuable a prize as the human genome; indeed Venter intended to give it away as a loss leader for Celera's human genome.

Despite the pretentious rhetoric in both sides' letters about humankind needing their cooperation, the public interest seemed rather better served by their fierce competition. Celera's mere entry into the race had advanced the consortium's projected delivery of most human genes from 2005 to the spring of 2000. If private capital existed to finance a parallel effort to the government's, how was the public worse off? If the human genome was so important, wouldn't two independent versions of it be even better than one?

The merger idea failed for personal and ideological reasons, and also because both sides could see a strong chance of trouncing the other. The consortium's production machine was swinging into full gear, operating round the clock and nearing the capacity to sequence 1,000 bases of DNA per second. It was only months away from its goal of sequencing 90 percent of the genome. Venter might claim to have cracked the fruit fly genome, but no one had seen his data and it might be full of large gaps; and even if his fruit fly genome was acceptable, with any luck he would stumble catastrophically on the human genome.

For Venter, success could not be guaranteed, but the fruit fly results looked good and there were features in the fruit fly genome's repetitive regions that suggested that the repetitive regions in the human genome should be even easier to sequence.

In the early months of 2000, both sides forged ahead independently, fighting an intense public relations campaign as their sequencing machines churned out genomic data. The consortium launched a devastating blow to Celera's stock price, inadvertently to be sure, when the Wellcome Trust, through the British government, persuaded Prime Minister Blair and President Clinton to issue a joint statement on March 14 that the genome was the property of mankind and "should be made freely available to scientists everywhere."

Given the somewhat unsettled state of patent law regarding genomic sequences, this innocuous sounding statement was misinterpreted by Wall Street as meaning that the White House was opposed to patents being issued on the human genome. Celera's stock price fell $39.75 to $149.25, even though patents were not a big part of its business plan. Other genomics companies fared even worse, and the slide spread to the rest of the NASDAQ market, knocking it for its second biggest point loss to that date. Fortunately for Celera, it had just raised $1 billion in February and had no pressing need to return to the capital markets.

On March 23, Venter unveiled the fruit fly genome. The venue, chosen for maximum impact, was the fruit fly biologists' annual meeting, held that year in Pittsburgh. Filing into the lecture hall, the 1,300 fly people, as *Drosophila* biologists call themselves, found on their seats a CD-ROM containing the fly's genome sequence. The president of their association proclaimed that they were "about to be handed an incredible tool that many of us only dreamed about for many years," whereupon Venter received a standing ovation.[33]

The applause was well deserved. The fruit fly genome appeared to be of very high quality. There were many small gaps, but none of any consequence. Like the consortium scientists who had sequenced human chromosome 22, Venter could not penetrate the fruit fly's centromeric DNA, a problem to be left for later, but it contained very few genes. Of the 2,783 *Drosophila* genes already found by fly people, all but five were present in the Celera-Rubin sequence, and the missing five may not be true *Drosophila* genes. And despite the ninety years during which biologists have studied the fruit fly, the genome sequence brought to light at one stroke a further 10,818 genes, underlining the enormous power of genomics to accelerate biological knowledge.

Publication of the fly genome proved that Celera's whole genome shotgun strategy worked on animal genomes and so was likely to succeed with the human genome. Collins rallied his troops by saying a week later that the consortium was two thirds of the way toward its goal and would complete its first draft of the genome by the end of June.[34]

As the finish date grew nearer, relations between the two sides continued to slide. At a tense hearing before a House committee on April 5, Collins, Waterston and Venter appeared as witnesses. Venter announced that his computer was about to start assembling the human genome and the sequence would be ready within three to six weeks, a forecast that in the event proved five weeks premature.[35] In case this remark did not sufficiently spook his competitors, he then launched into a public attack on the quality of the consortium's data, pointedly evening the score for

Maynard Olson's prediction before the same committee two years previously that Celera's data would be of "transient" value and "poor quality" and riddled with "over 100,000 'serious' gaps."[36]

It was also payback time against Collins who, also two years previously, had informed *USA Today* that Celera would produce the *"Mad magazine"* version of the human genome.[37] The genomic data the consortium was producing, Venter told the committee, were "an unordered collection of over 500,000 fragments" that was "nowhere close to being done." Referring to Collins's *Mad* magazine quote, Venter stated that Celera's goal had always been "to produce a high-quality human genome sequence that will stand the test of time."

Venter's feelings were not so hard to infer. As a little known scientist making his first contribution to genome sequencing, he had not merely been excluded from the establishment inner circle but publicly insulted as well, as when Watson had referred to Venter's brain gene sequences as the fruit of machines that could be "run by monkeys." Now that his fruit fly and human genomes were proving to be of high quality, it was time to fling back the stinging words.

Venter's testimony may well have given the consortium chills because it provided a helpful preview of what was likely to happen if both sides declared victory separately: Venter would trash the consortium's genome and repeat his criticism that its quality wasn't in the same class as his own. Such a charge might have merit. The consortium's interim version of the human genome—the go-for-the-genes, 90 percent version that Collins had promised for the spring of 2000—was not shaping up to be a thing of beauty. The consortium had not finished assembling many of its BACs, the 150,000-base-long chunks of DNA into which it had initially cut the genome. According to David Lipman, director of the NIH's National Center for Biotechnology Information and custodian of the GenBank database where the consortium posted its daily genome production, most of the consortium's data was in the form of short fragments about 10,000 bases long. The consortium knew which BAC each of the short fragments belonged to, and from its BAC map it knew where each BAC fitted to its parent chromosome. But often it did not yet know the correct order of the various fragments within each BAC. Thus its version of the genome was to a large extent unordered and in that sense, as Venter could not help pointing out, not a sequence at all.

The consortium could argue, as Lipman also confirmed was the case, that scientists could usefully search for human genes on its sequence, disjointed as it was. In this sense the consortium would have faithfully delivered on its promised goal of making almost all the human genes available. But to hold a big White House announcement of victory, as it was rumored the NIH had in mind, while Venter produced a visibly better genome and excoriated the consortium's, might prove exceedingly awkward.

So was this how the historic quest to decipher the genetic blueprint of humankind was to end up—not as a lofty triumph but a snarling cat-fight heard around the world? "The discovery and presentation of the human genome, one of the most important attributes of man, should be a time of great joy and happiness," said Norton Zinder of Rockefeller University. "For there to be all this vitriol and hatred just doesn't seem right." His colleague on Celera's scientific board, Richard Roberts, observed that the announcement of sequencing the human genome "should be a great celebration of humankind, not a race with a clear winner or loser."[38]

While the two rivals were assessing the complexities of the swamp their rivalry was leading them into, a new mediator stepped into the fray. Aristides Patrinos, the official who had long guided the Department of Energy's part of the Human Genome Project, had retained the confidence of both parties. He had amicably played a junior role to Collins in the government's Human Genome Project, even though the two agencies were formally equal partners. His department had given several contracts to TIGR, Venter's institute, for sequencing important microbes known as archea. Now Patrinos managed to bring the two proud, principled, mutually furious competitors together in a meeting in his house over beer and pizza.

This time there was no talk of complicated issues such as sharing electrophoretic traces. An elegant, minimalist deal emerged. Its terms seem to have been along the following lines: The two sides would make their victory announcements together and publish descriptions of their achievements simultaneously, though in separate articles and maybe separate journals. Neither would publicly criticize the quality of the other's work. Collins would extend to Venter the world's most prominent podium for his victory address: the White House.

And as balm for the consortium's collective insults of the past, Venter would be praised to the heavens by a president, a prime minister, and his principal opponent.

Announcement of the pact caught almost everyone else by surprise. Just five days later, on June 26, 2000, the principals of the two teams gathered in the White House, all save Sulston and the British team, who were having their separate events in London, and Michael Hunkapiller, the prime architect of Celera, whom fate for some capricious reason had sidelined with chicken pox.

For Watson, the architect of the consortium, an amazing cycle of his life had closed. Playing with cardboard cutouts of the four DNA bases on a tiny desk in the Cavendish Laboratory at Cambridge forty-seven years before, he had been the first to see the secret of DNA's incredibly beautiful structure. Now, once again seeing further and deeper than his fellow biologists, he had with great effort drawn another amazing treasure, the human genome, into the realm of comprehension. But this triumph was not the same. The discovery with Crick had been a true matching of minds, a shared achievement that each acknowledged could not have been made without the other. Sequencing the genome had been an altogether different experience. But how could the complexities of the eight-year struggle to bring the genome to light be expressed in a few simple sentences? When the applause had ebbed and reporters pushed through the crowd to ask what he thought, he escaped them as fast as possible.

The White House ceremony secured by Patrinos's mediation did not mark a treaty between the two camps, merely a temporary and expedient truce. Personal relations were soon back to normal. When the consortium announced it was setting up a separate alliance to sequence the mouse genome, known as the Mouse Sequencing Consortium, Venter accused its members of duplicating his work and wasting public money. He suggested that the mouse work would distract the consortium from the task of completing its human genome. "The worry we always had was that they would not finish it," he said.[39] The barbed reference was to the consortium's frequent comment that it was going to produce the true archival version of the genome, not the hasty commercial job to be expected of Celera.

The political agendas of the two sides remained as separate as ever. The public consortium made every effort to ensure that academic biologists would have access not just to the genome sequence but to the several other vital tools of the genomic era. The NIH and the Wellcome Trust spearheaded this effort, joined by large pharmaceutical companies such as Merck and SmithKline Beecham, which had no desire to be at the mercy of Celera and other genomics companies or to see academic biology wither away as genomics disappeared behind commercial walls. Both companies supported the Mouse Sequencing Consortium. Merck had previously sponsored a public EST database to compete with those of Human Genome Sciences and Incyte. In April 1999, ten pharmaceutical companies and the Wellcome Trust set up the SNP Consortium, an effort to describe and make freely available the single nucleotide changes, known as SNPs, pronounced "snips," which are the principal source of human genetic variation. Celera too had a large SNP program that included its own SNPs and those available from public databases.

After the White House announcement, the two teams returned to their laboratories and the real battle, that of preparing their respective genome sequences for publication by trying to figure out where the human genes were hidden and how many were listed in the genome's instruction manual.

3
The Meaning of the Life Script

Though the two sides' claim to have sequenced the human genome sparked headlines around the world, they were celebrating victory before the battle. Each team now had to interpret its genome sequence by describing its major features and locating the gene. From the quality of the respective reports a clear winner could emerge in the eyes of their fellow scientists.

This first analysis of the human genome was just as much a landmark in scientific history as the years spent decoding the sequence. But it was compressed into a few months' whirlwind of activity as the two sides set about interpreting the strange and enigmatic script they had wrested from the human cell. Talk of a joint annotation conference at which the two sides would compare notes in interpreting the human genome soon evaporated. Celera and the consortium worked with different groups of experts and published their reports in rival scientific journals, Celera choosing *Science* in Washington, D.C., and the consortium *Nature* in London. The sharp elbowing continued to the bitter end, with academic biologists including Eric Lander, chief author of the consortium's genome report, lobbying hard to persuade the editor of *Science* not to accept their rival's paper except under terms unacceptable to Celera.

The academics demanded that Celera make its genome sequence

freely available through GenBank, even though GenBank could not meet Celera's condition that its commercial rivals be prevented from downloading Celera's data and reselling it. The editor of *Science* allowed Celera to be the custodian of its own data on condition that any part of it be made freely available to scientists for checking, a decision to which the academic biologists objected. Venter felt their real purpose was to deny him the prestige of being published in a leading academic journal.

Well before the White House announcement, both sides had started preparing to analyze their respective versions of the genome. Venter had more or less founded the art of genome interpretation when he published the first genome of a bacterium in 1995. He had learned then an important lesson: the best way of interpreting a genome is to compare it with the genome of a similar organism. He had long since decided that the mouse's genome would be a critical tool for interpreting the human genome, because comparison of the two long ago mammalian cousins would reveal by their regions of DNA sequence similarity all the features that nature had found it necessary to conserve. He had risked switching his sequencing machines from human to mouse DNA at the earliest possible moment so as to have both genomes in hand for the task of locating the human genes.

Another advantage for Celera was that its version of the human genome was much less bitty than the consortium's. Celera's vast computer, the largest in civilian use, had assembled 27 million of the 500-base pieces analyzed by the sequencing machines into long, mostly continuous scaffolds that straddled the genome. The consortium's genome was divided into thousands of the small sub-jigsaws known as BACs, chunks of DNA about 150,000 bases in length. The BACs had been completed for the two shortest human chromosomes, numbers 21 and 22, but over most of the rest of the genome were still in small pieces, many 10,000 bases or so in length. It was possible to hunt through these fragments for genes, but not at all easy. The consortium had not tried to assemble them by computer because it did not see the need to do so. Robert Waterston, director of the sequencing center at

Washington University in Saint Louis, had prepared a BAC map that showed how one BAC overlapped another across the genome in a complete tiling path. With this BAC map in hand, the same method by which Waterston and John Sulston at the Sanger Centre had sequenced the roundworm's genome, there seemed no need to invest in the complex computing and assembly programs that were a necessary part of Celera's strategy.

Though Sulston and Waterston had laid the scientific groundwork for the consortium's sequencing effort, it was Eric Lander, director of the Whitehead Institute's sequencing center and a mathematician by training, who took the lead in analyzing the genome. In December 1999 he started to invite computational biologists—a new discipline devoted to computer analysis of genomes—to join in a genome analysis group. The group had no government funding, according to Lander, and did its work mostly by phone and e-mail.

Meanwhile Venter, who had convened an "annotation jamboree" of outside experts to help find the genes in the fruit fly genome, decided there was now enough expertise within Celera to undertake the first analysis of the human genome in-house, with the help of a few consultants.

The consortium might have been hopelessly outgunned in the interpretation phase of the genome race had it not been for a chance encounter, although one made possible by the consortium's open nature. One of the computational biologists approached by Lander in December 1999 was David Haussler of the University of California, Santa Cruz, whom Lander invited to help locate the genes. Haussler decided that before looking for genes, it would be best to put some order into the jumble of fragments within each BAC. He believed there was enough information, some created inadvertently by the consortium and some from other sources, for an assembly program to order and orient the intra-BAC fragments, and he at once started writing such a program.

To create the computing facility to run the program, he persuaded his university chancellor to advance him the money for a network of one hundred Pentium III computers. But the programming went slowly. In May, when a graduate student of his e-mailed to ask how the genome assembly program was going, Haussler replied that things were looking grim.

The student, James Kent, then offered to write an assembly program himself, using a simpler strategy. Haussler replied, "Godspeed." Four weeks later, Kent had completed an assembly program that in his supervisor's opinion might have taken a team of five or ten programmers six months or a year. "He had to ice his wrists at night because of the fury with which he created this extraordinarily complex piece of code," Haussler said of his student.[1] Kent, who had previously run a computer animation company before returning to school to study computational biology, first used the program to assemble and order all the pieces in the consortium's genome on June 22, 2000. In doing so he gained a three-day lead on Celera, whose assembly program had encountered unexpected problems. Venter completed his first assembly of the human genome on June 25, just the night before the White House press conference.[2]

When Celera and the consortium published their analyses of the genome in February 2001, it was clear that the consortium's rested heavily on Kent's improvised assembly program and the computer network put together by his supervisor. Venter was astonished that his competitors had at the last minute managed to extract so much sense from a genome sequence that in his view had been so hopelessly chaotic. "They used every piece of information available," he said. "It was really quite clever, given the quality of their data. So honestly, we are impressed. We were truly amazed, because we predicted, based on their raw data, that it would be nonassemblable. So what Haussler did was, he came in and saved them. Haussler put it all together."[3]

In their first glimpse of the human genome, the two teams came to similar conclusions, of which the most surprising was the far smaller than expected number of human genes. Both found about 30,000 protein-coding human genes, far fewer than the 100,000 human genes that textbooks had estimated for many years. The 100,000 number had seemed especially credible after the *C. elegans* roundworm was found to have 19,098 genes and the fruit fly 13,601.

Celera said its new gene-finding program, named Otto, had predicted 26,588 human genes for sure, with another 12,731 possible genes. The consortium estimated the human instruction set at 30,000 to 40,000 genes. Both sides favor a number at the lower end of their respective estimates because gene-finding programs tend to overpredict, and 30,000 genes seems for the moment the preferred figure.[4]

These first readings of the human life script, however awesome for biologists, were a little baffling for the lay audience. All this effort, just to find that people have only 50 percent more genes than a worm? Yet the human instruction manual was never likely to yield up all its secrets at first glance. It was hardly surprising that the first scan of its pages should produce more perplexity than enlightenment.

The human genome is written in an ancient and vastly alien language. It is designed for the cell to use, not for human eyes to make sense of. Its four letter alphabet, represented as A, T, C, and G for the four different bases of DNA, is so hard to parse that there would be no point in printing out the whole genome sequence. If anyone were to undertake so futile an effort, the result would occupy three hundred volumes the size of those in *Encyclopaedia Britannica*, each page of which would carry an almost identical looking block of letters, unbroken by spaces, punctuation, or headings. Nature's only subdivision of the genomic script is to package it in twenty-three chapters, the chromosomes, each of which is an enormous DNA molecule festooned with the special proteins that control it. These range in size from chromosome 1, a blockbuster 282 million bases in length, to chromosome 21 at a mere 45 million bases.

The challenge of the task that faced Celera and the consortium was to decipher the 3 billion letters in the script with machines that could read only five hundred letters at a time, losing the position of the fragment as they did so. The order of the five-hundred-letter pieces had to be reconstructed largely by creating so many that they overlapped, and so that through the overlaps the original chromosomal sequence could be inferred. Moreover both teams held themselves to an eventual accuracy standard of less than one error per 10,000 bases.

The interpretation of the genomic script is likely to prove as hard as the sequencing. So far just the major features have been recognized; doubtless many details remain invisible because of biologists' still substantial ignorance as to how the genome works.

One major feature of the genome's geography, which has prevented both Celera and the consortium from estimating its exact size, is the centromere. This is a stretch of DNA, more or less in the center of each chromosome, which is recognized at the time of cell division by the machinery that pulls duplicated DNA strands apart so that each daughter cell can receive a full set. The centromere consists of the same sequence of bases repeated many times over. Along with certain other problematic regions of the chromosomes, known collectively as heterochromatin, the centromeres cannot be sequenced by present day methods. Since they contain few or no genes, the anonymity of their DNA probably does not much matter.

Including its heterochromatic regions, the human genome now seems to be 3.1 billion bases in length. The rest of the DNA, known as euchromatin, is 2.91 billion bases in length, according to Celera.

Of the 2.91 billion bases, one quarter contains genes but the other three quarters of the vast terrain, as far as can be seen at present, is a graveyard of fossilized DNA, evolutionary experiments that didn't work, and dead genes on the road to extinction. The original name for this non-gene DNA—junk DNA—has been regarded as presumptuous because it assumed, without proof, that the so-called junk was useless. But on first glance, much of the non-gene regions of the genome are indeed full of junk.

The principal occupants of these regions are rogue pieces of DNA that have been able to copy themselves and insert the copy elsewhere in the genome. Called mobile DNA or transposons, these parasitic elements seem in some cases to have been derived from working genes and taken on a life of their own. The copying seems to serve no purpose other than cluttering up the genome. The consortium reported finding more than 850,000 LINEs and 1,500,000 SINEs, as the two largest families of rogue DNA are called. (The acronyms stand respectively for long and short interspersed nuclear elements.) A LINE is about 7,000 bases in length and a SINE about 200 bases. These and the two other families occupy a total of 1.229 billion bases of DNA, or more than one third of the genome.

The good news about the transposons is that most are dead, in the sense that they ceased to copy themselves thousands of years ago. They clock up mutations, just like any other segment of DNA, and the older they are the more mutations they have. By use of this mutational clock, the consortium determined that only one family of LINEs is still active in the genome, together with a SINE family that uses the LINEs' copying mechanism. Transposons are a genomic hazard, though not a large one, because the copies they make of themselves are inserted back into the genome at random sites. If the insertion occurs in the middle of a gene, the gene is likely to be disrupted. The active LINE element was first noticed because it had disrupted the gene for the blood-clotting Factor VIII in a hemophilia patient. Humans, with most of their transposons long ago placed on fossil status, are for unknown reasons much better off in this respect than the mouse, in whose genome transposons are still vigorously copying themselves and cluttering up the mouse's genetic patrimony.

The Celera team calculates that the region of the genome devoted to human genes occupies just a quarter of the euchromatic DNA, with the average gene sprawling over 27,000 bases of DNA. Genes consist of alternating stretches of DNA known as introns and exons, an arrangement that now seems particularly important in explaining how human complexity is generated with so few genes. The intron-exon system is the basis of a baroque system known as alternative splicing, which works as follows. A gene is activated by a set of special proteins that assemble on its control region, just upstream of the exon-intron mosaic. This transcription complex, as it is called, moves down the double strand of DNA, pushing one strand out of the way and making a copy of the other.

The material of the copy is not DNA but its close chemical cousin, RNA or ribonucleic acid, so called because its nucleotides are composed of the chemical unit known as ribose in place of DNA's deoxyribose. These RNA transcripts are then edited by a dexterous piece of cellular machinery known as a spliceosome, which snips out the introns and splices the exons together in a much shorter transcript. The string of united exons is then exported from the nucleus of the cell to the ribosomes, the protein-making machines in the cell's periphery or cytoplasm.

As the edited RNA transcript ratchets through the ribosome, the order of its bases dictates the composition of the protein chain that is assembled in lockstep with its passage. With each three bases of the RNA transcript, one amino acid—the chemical units of which proteins are made—is added to the growing protein chain. Each triplet of bases codes for one of the twenty types of amino acid although, since there are sixty-four possible combinations, some triplets code for the same amino acid, and three triplets also serve as a stop sign. This all-important relationship between the DNA/RNA world and proteins is known as the genetic code and must have been established at the dawn of life on Earth some 3.8 billion years ago.

Though proteins are made as a linear sequence, the chain folds up into a very specific three-dimensional structure dictated by the order of its amino acids. The part of each amino acid that forms the protein's backbone is standard, but each has a chemically different side group that juts out from the backbone, and it is the combination of these different groups that gives protein molecules their remarkable versatility. Some proteins serve as structural materials, like the stretchy collagen fibers of the skin, some as enzymes that run the cell's metabolic reactions, and some as complex machine tools, like the amazing topoisomerases that specialize in unknotting DNA when it gets in a tangle.

A neat feature in nature's design of proteins is that they have a modular design. A single protein can contain several modules or domains, each of which performs a different function. Some of the central roles in human cells are played by single proteins with a large number of domains, each of which regulates a complex circuitry of lower level proteins.

This is where the intron-exon structure of genes comes into play.

When the spliceosome processes the RNA transcript of a gene, it can produce alternative editions, often by skipping exons, occasionally by including introns. The result is that a single gene can produce a family of different proteins, each with a different set of domains and different overall properties.

Biologists have no clear idea yet as to how cells control alternative splicing, as the process is called. In some cases different control regions in front of the gene may be selected by the transcription complex. Or the introns themselves may control which exons get edited out. Sometimes different splice forms are produced in different types of cells. The dystrophin gene, for example, one of the largest in the genome, occupies over 2.4 million bases of DNA that contain seventy-nine exons. It takes the cell sixteen hours just to make a transcript. The spliceosomes discard 99 percent of the transcript, but the edited version generates a nonetheless vast protein of 3,685 amino acids. Mutations that occur at various sites in the gene can impair the protein and produce the spectrum of diseases known as muscular dystrophy, which is how the protein was discovered and named. Though the manufacture of dystrophin seems an inefficient process, the body makes double use of the gene. The dystrophin gene is also activated in brain cells, but there it is alternatively spliced. Brain cells use only a subset of the seventy-nine exons and produce a much smaller protein.

Alternative splicing, though its mysteries are only just beginning to be probed, may well hold one of the sources of mammalian complexity. Up to 60 percent of human genes may have alternative splice forms. This may enable human cells to generate perhaps five times as many proteins as the worm or fly, despite having only twice the number of genes as the fly, the consortium said in its analysis of the human genome.

Another source of mammals' complexity seems to lie in the more sophisticated architecture of their proteins, which tend to have more domains than those of the fly and worm. Most protein domains are very ancient. Only 7 percent of the domains in the human proteome (that's the genomic age word for all the proteins made by a genome) are not found in lower animals, showing that invention of new domains was not so important in the design of mammals. The difference is that human proteins tend to have more domains, which makes possible not just more complex proteins but a much richer combination of interaction among proteins.

"The main invention seems to have been cobbling things together to make a multi-tasked protein," said Collins, the consortium's leader. "Maybe evolution designed most of the basic folds that proteins could use a long time ago, and the major advances in the last 400 million years have been to figure out how to shuffle those in interesting ways. That gives another reason not to panic," he said, referring to the unexpectedly small number of human genes his team had found.[5]

There's another possible explanation for the small number of human genes found by Celera and the consortium, which is that both teams seriously undercounted. This is the belief of Venter's former colleague William Haseltine. Haseltine, chief executive of Human Genome Sciences, had long pegged the number of human genes at 120,000 or more, as had Randal Scott, chairman of Incyte. In fact Scott's prediction of 142,634 genes, made in September 1999, was one of the highest on record. In the wake of the new analysis, Scott said he accepted the new lower tally. But Haseltine stood firm with his high estimate, even though he was now in a minority of one against almost all the world's leading genome analysts.

Haseltine's rationale was hard to assess, since he had not published any of his findings, but also hard to dismiss. He believed that Venter and the gene finders had gotten it all wrong. To sequence the genome, to identify its genes, to discover in which cells the genes were turned on—was all, in his view, a prodigious waste of effort when there existed a far simpler and surer method of identifying the human gene repertoire. The shortcut, pursued of course by Human Genome Sciences, is to let the cell read out the baffling information in the genome and then capture the cell's RNA transcripts. Human Genome Sciences had invested enormous effort in capturing the genes made by different types of human cells, including those of fetal tissues in various stages of development.

The method of capturing RNA transcripts had been exploited first by

Venter and was the basis of his original partnership with Haseltine. The transcripts, which are usually sequenced only in part, are called expressed sequence tags, or ESTs, and are extremely useful for locating the genes from which they are copied. But EST collections are also known to overpredict the number of genes from which they are derived, in part because of alternative splicing and other vagaries of the spliceosome's operation. It was in large measure through recognizing that ESTs pointed to too high a number of genes that the consortium and Celera had arrived at the low number of 30,000.

But Haseltine said he was confident that he had removed the alternative splice forms and other known sources of confusion from his EST collection. With the resources of Human Genome Sciences, he had been able to determine the full length sequence of 90,000 ESTs, which he believed represented 90,000 different genes.

When consortium biologists published the sequence of human chromosome 22 in December 1999 they reported that they had identified at least 545 genes in it. But Haseltine figured he could see twice that number of his genes located to chromosome 22; in other words, the best conventional gene finding methods had picked up only one gene in two.

"No new discoveries were made, no new genes were found, and the authors go to great length to tell us that chromosome sequence cannot be used to find genes. I call that the biggest untold secret of the Human Genome Project," he said.[6]

Warming to the same theme a few months later, Haseltine observed that "People who sequence DNA are the least likely to know how many genes there are."[7] Later, when almost everyone who did sequence DNA had concluded that there were only 30,000 human genes, Haseltine argued that they had erred en masse because both teams had used the same methods of gene finding. The methods are admittedly imperfect and can overlook genes. One kind of evidence the present gene-finding programs rely on is a homology search, meaning that the programs search the DNA databases for any genes from other organisms that have a DNA sequence similar to any in the human genome. Any human DNA sequence that is homologous with, or similar to, a known gene in another genome is likely to be a gene itself. This is a powerful search method but will miss any gene that is so far unique to humans and has not yet been catalogued in the databases.

Unfazed by being in a minority of one, Haseltine predicted that the number of known human genes would steadily rise to meet his projections of 100,000 to 120,000. Venter, though he disagreed with his former colleague's number, wrote in his analysis of the human genome that ultimately the only way to determine the number of genes would be to capture the genes made in different types of cells. This is indeed the approach Haseltine has taken.

The first analyses of the human genome have brought home the long familiar fact that all organisms are intimately related to one another through being twigs on the same tree of life. But even evolutionary biologists may have been surprised by the overwhelming degree of similarity of people to other forms of life at the DNA level. About the only thing people have in common with the mouse is that we are fellow mammals, although separated by 100 million years from our last common ancestor. Yet Venter, having assembled the mouse genome, said that of the 26,000 confirmed human genes he could find only 300 that had no counterpart in the mouse. On this basis he expected the chimpanzee, our closest living relative, to have essentially the same set of genes, with the difference between the two species being caused by variant versions of the same genes.

The consortium, for its part, asserted that at least 100 human genes seem to have been borrowed from bacteria, presumably via some ancient infection. But this conclusion was quickly shot down. Its basis was heavily criticized by scientists at the Institute for Genomic Research and elsewhere, and Lander, the lead author of the consortium's paper, did little to defend it.[8]

The similarity between the human and mouse genomes shows how far biologists are from being able to make the link between a genome and the organism that is based on it. Evidently small and subtle changes at the genomic level produce enormous differences at the level of the whole organism. A vast amount of research lies ahead before the human mechanism is fully understood in terms of its genetic instruction manual.

One necessary step is to continue the work that has begun on compiling a full catalog of human genes. Present gene-finding programs are powerful but not very accurate. Essentially, they look for "open reading frames," stretches of DNA that start with ATG, the triplet of bases used to initiate a protein chain, and continue for a plausible length without any of the stop-sign triplets. This method works well in bacteria, which have very compact genomes, but gene-finding programs get confused by the intron-exon structure of animal and plant genes. The junction between introns and exons is not well defined, and some of the cues the spliceosome uses for stripping out the introns are not yet understood. Nor are the rules that govern alternative splicing. Since the theoretical basis for detecting gene sequences in the genome is still incomplete, biologists supplement their prediction programs with empirical data. These include the DNA sequences inferred from known human proteins, which must for sure be reflected somewhere in the human genome; EST sequences—the snippets of RNA transcripts captured from living human cells; homology searches; and a direct comparison of the human with the mouse genome. The combination of all these data helps predict the exons in the human genome that may be parts of human genes.

The human genome is more than a mere list of protein parts; it also embodies the program for the operation of the human cell. The program lies in control sequences of DNA that are placed upstream of the genes and are recognized by the gene transcription complexes. Molecular biologists have identified many of these control sequences, but there are doubtless many more to be discovered. The mouse genome is again of help. When the mouse genome, suitably reorganized so as to correspond to the somewhat different arrangement of human chromosomes, is laid alongside the human genome, the regions of conserved DNA sequence that do not correspond to exons may be control sequences.

Besides exons and control sequences, there is a third category of similarities between the mouse and human genome, Venter has said. Its nature remains unknown, though any DNA sequence that nature has found worth conserving for 100 million years must be important. One possibility is that there are many more RNA-making genes to be discovered. Most genes make proteins, but many of the cell's most vital pieces of machinery are made largely of RNA molecules. These include the spliceosome that edits RNA transcripts and the ribosome that translates the transcripts into proteins. RNA molecules are known to perform certain other roles, such as silencing one of the two X chromosomes in a woman's cells, so that the dosage of genes from the X will be the same as in a man's cells. RNA molecules can twist up into elaborate 3-D shapes, as do proteins, and can also catalyze chemical reactions. Given nature's propensity for using whatever is at hand, the full extent of RNA genes may not yet be known.

While computational biologists continue to refine their identification of human genes, a new branch of biology called proteomics has emerged. In one sense proteomics is just the study of proteins on a genome-wide scale, giving plain protein chemists the chance to style themselves proteomicists. But ingenious new methods are enabling biologists to study all or many of the proteins from a cell en masse. One technique, involving advanced mass spectrometry, allows all the proteins to be identified. Another, known as the yeast two hybrid system, shows which proteins interact with one another in the cell, an important first step to deciding what an unknown protein does.

Study of the genome and its proteins lays the groundwork for understanding the living human cell, which in turn is the basis for understanding all human disease. Although enormous progress has been made since the beginning of molecular biology in 1953, many essential features of the mechanism are only dimly understood. The human body is thought to contain around 100 trillion cells. All are the progeny of a single fertilized egg but each, by some still mysterious alchemy, has morphed into its specialist adult role.

There are about 260 different known types of human cell, with doubtless more to be discovered.[9] All these cells share the same genome but must make use of it in different ways. Most of the genes must be permanently switched off, or the cell would be in chaos. Presumably each type of cell uses a common set of house-keeping genes and a suite of genes reserved for its use alone. But no one yet knows how to analyze the genome in terms of which sections are designated for use by a kidney cell and which by cells of skin or lung or liver. Nor is it yet clear how in the process of development each type of cell is assigned its own character and pattern of gene expression.

Most of the body's cells are probably in a state of intense molecular activity. The proteins of a cell are constantly interacting with one another, in effect performing complex calculations, the outcome of which is a decision sent to the nucleus to turn a gene on or off. Messages from both neighboring and faraway cells continually arrive at the cell's surface, bearing instructions that the receptor proteins in the cell's outer membrane convey to proteins in the interior and thence to the nucleus. Inside the nucleus, as the result of all the internal calculations and external messages, an array of transcription factors is copying genes, maybe hundreds or thousands every second.

This elaborate activity all takes place within a minute space. Imagine the smallest speck of dust you can see. This speck is the size of five average-sized human cells. These minuscule corpuscles are the protean clay of which the body is sculpted.

Within the cell are various compartments, of which the most prominent is the nucleus, which occupies a mere 10 percent of the cell's volume. The nucleus is the protected residence of the genome. In the rest of the cell, a fluid-filled compartment known as the cytoplasm, are other specialized structures such as the ribosomes, which manufacture the cell's proteins, and the mitochondria, the cell's energy production units. The mitochondria were once free-living bacteria that billions of years ago were captured and enslaved by animal cells; the mitochondria have their own, much degraded genome, a little circle of DNA containing a mere 16,569 nucleotide pairs.

Inside the nucleus the genome is packaged in the form of twenty-three pairs of chromosomes, each consisting of a single giant DNA molecule wrapped in the special proteins that protect and manage it. Despite the minuscule volume of the cell nucleus, the chromosomes are sizable objects. If fully stretched out, chromosome 1, the longest, would measure about 8.5 centimeters (3.35 inches).[10] The forty-six chromosomes in the nucleus would stretch for just over 7 feet if laid end to end. It is an extraordinary feat of engineering for nature to have packed a 7-foot tape into so tiny a volume, yet still allow the cell unfettered access to all the parts it needs. "DNA stores 10^{11} gigabytes per cubic centimeter—it's almost the greatest molecular packing density one could expect to get at the molecular level," says Randal W. Scott, chairman of Incyte Genomics.[11]

Though it will be the work of decades to understand this miniature miracle of biological computing and construction, the process of translating knowledge about the genome into medical advances need not wait and has indeed already begun.

So who was the winner of the great race to sequence the human genome? Venter had accomplished much of what he set out to do, although, as discussed below, there was a serious inadequacy in the *Science* article describing his interpretation of the genome. He produced a very useful, though not complete, version of the human genome by February 2001, the date when most scientists could get access to at least parts of it. Without Venter's fierce competition, the consortium might well have continued on its original trajectory of providing a complete genome (complete, that is, apart from the heterochromatin DNA, which cannot at present be sequenced) by 2005, Watson's original target date.

Venter had not only brought forward the availability of the human genome by four years, he had also sequenced and assembled the mouse genome, an invaluable aid to interpreting the human genome. All in all, it was a spectacular achievement that validated the extraordinary risks that he and Michael Hunkapiller had taken in designing the project and that Tony White, chief executive officer of Applera, had taken in backing it.

But the consortium had also succeeded. Its draft version of the genome, published at the same time as Celera's, was surprisingly comparable even though Venter had been able to use both his own data and the consortium's whereas the consortium had only had its own. The consortium could thus claim a share of Venter's credit, although it had borrowed some of Venter's methods, such as the use of paired-end reads (having the sequencing machines read both ends of DNA fragments of known length).

The consortium bore the organizational burden of being spread among centers in six countries, although the brunt of its effort was born by John Sulston at the Sanger Centre near Cambridge in England and Robert Waterston's center at Washington University, Saint Louis, later joined by Eric Lander at the Whitehead Institute in Cambridge, Massachusetts. It was also dependent on, and at times held back by, the uneven flow of government funding. Despite all these impediments, the consortium was an outstanding technical success that remained on schedule and within budget, not so common a record for a government project.

The high noon duel with Venter was in any case subsidiary to the consortium's final goal, that of producing the entire genome sequence by the revised target date of 2003. Both sides' versions of the genome contained numerous gaps, although these were mostly in the regions of repetitive DNA and probably contained few genes. The consortium had committed itself to close every gap, except those in the heterochromatin, by 2003. Although Venter too planned to improve his sequence, he had not made the same commitment to completeness.

Initial reactions from some researchers who had examined both genome versions suggested that Celera's was "more accurate, easier to read and more complete than the rival version."[12] With twice as much sequence data to draw on as the consortium, that was not surprising. And Celera desperately needed a sequence that was in some way better than its rival's, which was available for free. Venter's secret weapon was his mouse genome sequence, available only to his subscribers, but the consortium was working busily to prepare its own mouse sequence, and its members determined that not only the genome but every important interpretive tool should be freely available for all. The consortium was also busily working to improve its own human genome sequence. With two moving targets, judgments as to a winner were difficult and likely to prove temporary.

The inadequacy in Celera's *Science* article, which consortium scientists were quick to seize on, lay in Venter's strange neglect of his own shotgun method. He had his computer assemble the human genome by two different methods. One was the whole shotgun assembly and the other was a hybrid method that relied not only on the consortium's data but also on knowledge of the BACs' positions on their chromosomes, a borrowing that took direct advantage of the consortium's method. With two versions of the genome in hand, Venter then ignored the shotgun genome and based all his further gene identification and genome analysis on the hybrid version genome.

The consortium's scientists, once they had digested Celera's paper, were furiously contemptuous of it. Venter had claimed all along, with much swagger and braggadocio, that he would produce the better version of the human genome, yet the version he preferred in his *Science* paper leaned heavily not only on the consortium's genome data but also on its method. In the consortium's view, that was tantamount to someone downloading the consortium's work, adding a little DNA data of his own, and claiming the result as his own superior product.

Even the genome version Celera prepared by its whole genome shotgun method was not wholly independent of the consortium. Because of Venter's decision to switch his sequencing machines over to the mouse genome at the earliest possible moment, he had far less human data than originally planned and needed to borrow data from the consortium. But the data the consortium made publicly available in GenBank was partially assembled. Gene Myers, Celera's software designer, had previously asked the consortium for its raw reads of the *C. elegans* genome but had been rebuffed. Figuring the consortium centers were unlikely to give him their human reads either, Myers decided to download the assembled human data and artificially shred it in his computer into 500-base-length fragments.

The shredding would remove the many misassemblies he suspected existed in the consortium's genome. But he needed the artificial 500-base reads to assemble themselves across any gap in his own shotgun assembly. So he shredded the data twice, with the second set of shreds overlapping the first by exactly half a length.

The consortium biologists now had a powerful stick with which to beat Celera. Led by Lander, they charged that their shredded data would have reassembled itself in Celera's computer, injecting the positional information of the BAC-by-BAC method into Celera's whole genome shotgun, and that the shotgun would have failed without it.

"The WGS was a flop. No ifs ands or buts," Lander wrote in an e-mail referring to the whole genome shotgun. "Celera did not independently produce a sequence of the genome at all. It rode piggyback on the HGP. I have no objection to their using the data for their commercial databases, but it does seem odd to publish a paper reporting a sequence of the genome based largely (>60%) on other people's data."[13]

Phil Green, a computational biologist at the University of Washington, Seattle, and author of Phred and Phrap, two widely used genome programs, had arrived independently at the same conclusion. "I think basically they could not have done the human genome using a whole genome shotgun and I think they realized that at some point, which is why they depended so heavily on the public data," he said.[14]

Olson, who had forecast catastrophic problems for the whole genome shotgun in June 1998, maintained he had been right all along. He had predicted 100,000 serious gaps, and indeed there were 120,000 gaps in the shotgun version of Celera's genome. "Celera wanted to have it both ways," he said. "There is no question you get more data quicker with a shotgun, but you do it at the price of ending up with a mess on your hands that is unmanageable. Venter in June 1998 claimed that this was not a quick and dirty approach, that it would produce a sequence that met or exceeded best current standards. That claim was absurd at the time and remains absurd."[15]

The refrain was taken up by Collins, the consortium's leader. "Careful analysis leads most observers I have consulted to conclude that a pure whole genome shotgun approach is unworkable for a genome of the size and repeat density (50%) of the human," he wrote.[16]

Venter was outraged at his critics' attacks which, despite all previous experience to the contrary, he and his colleagues seem not to have expected. To have assembled two versions of the genome, both dependent to different degrees on the consortium's data, and then to have chosen the more dependent version, was not the most brilliantly conceived strategy for persuading his rivals to cry uncle. Indeed it positively invited them to cry foul.

His choice had been shaped by the competing needs of his commercial and academic goals. To sign up clients for Celera's database, he needed the best possible genome sequence. He explained that since the hybrid version of the genome contained about 2 percent more of the DNA sequence, he decided to analyze that instead of the shotgun version.

In that case, a totally independent shotgun version of the genome would have seemed essential for an article in the scientific literature written in explicit competition with the consortium's. "We didn't anticipate that anyone would say it didn't work," Myers said when asked why he didn't prepare a shotgun version of the genome using only Celera's data. "It didn't cross my mind. All we would have gotten would be to prove a point. It seemed silly to waste a lot of money on a point of pride, that's all there is to it."[17]

Myers says he did not use any positional information from the shredded data, which reassembled itself only across gaps determined by the scaffolds in the shotgun. If so, there is probably little merit to Lander's charge that the whole genome shotgun would have failed without the public data; it would just have been a little less complete.

Myers and Venter also note that they have assembled the mouse genome, which is very similar in size and structure to the human genome, by the shotgun approach and using only Celera's data. The results with the mouse genome are very similar to the shotgun version of the human genome—further proof, they say, that the shotgun worked.

To make certain, though, they started a 20,000-hour computer run to assemble the human genome using only Celera's data—just to prove the point they would have been wise to prove the first time around.

The shotgun method was a clear success with the fruit fly genome; but whether it worked better than the BAC-by-BAC method for the larger and more complex human genome may not be clear for a long time, if ever. Much depends on the yardstick applied. For completing the last possible base in the human genome, the consortium's method will probably be better: Celera is not even trying to settle the position of every base. But Celera's combined mouse and human database may prove a winner.

Should the Nobel Foundation's medical committee decide to award its prize for the human genome, it will have a hard choice. The Swedish jurors usually shy away from messy situations where credit is disputed. But there would be a neat way for them to sidestep the issue. Their prize, which can be awarded to a maximum of three people, could be split between Venter, Sulston and Waterston on the grounds that Venter sequenced the *Haemophilus* genome, a feat that revolutionized microbiology, while Sulston and Waterston laid the basis for sequencing animal and human genomes by completing the genome of *C. elegans*. And those three probably contributed most to sequencing the human genome, although the drama included several other important players, starting with Fred Sanger.

Prizes and wrangling aside, both sides, it could be said, had accomplished extraordinary feats. Each set a high goal and achieved it, with their competition serving the public interest. In such a contest, there could be no losers.

4
To Close Pandora's Box

The race to sequence and interpret the human genome has culminated in an unparalleled gift to biomedical researchers and physicians. With the genome sequence in hand, they can hope to understand all human disease at the genetic level and, in time, to develop treatments or cures on the basis of this understanding. At least in principle the prospect of conquering almost all disease—the dream of closing Pandora's box—can now be entertained.

Over the next decade, a cascade of genome-driven advances is likely to enter medicine, providing a raft of new drugs, sophisticated genome-based tests, and novel methods of treatment. Many of the drugs will be new human proteins discovered in the genome or adaptations of them. The use of new proteins to manipulate cells may help open out a novel branch of medicine, in which patients would be treated with specially treated cells to replace diseased tissues. This technique, discussed further in the next chapter, is referred to as cell therapy or regenerative medicine.

Perhaps the advance most noticeable to patients, and the soonest to become widespread clinical practice, will be the increasing use of genome-based diagnostic tests. In individualized medicine, as it has been called, more and more drugs will be packaged with diagnostic tests designed to identify the patients who will respond best and to screen out those likely to suffer side effects. And as the genetic basis of more diseases is understood, it will be easy to devise tests for susceptibility to each, although treatments will take longer to develop. Many tests could be useful even without an accompanying treatment, if they point to some preventive change in lifestyle. Knowledge of a genetic susceptibility to heart disease, for example, might help a person pay particular regard to health advice on diet and exercise.

The medicine of the future may include a scan of a patient's entire genome, looking for all variant genes that predispose toward disease. In a minor way, universal genetic testing has already begun, although for only a handful of diseases. When a newborn in the United States, Japan and other countries is a tender twenty-four hours old, it gets a first taste of medical procedures: a nurse jabs a small needle into the side of the heel and lets a few drops of blood fall onto a piece of filter paper. The blood is then tested, principally for phenylketonuria, which affects 1 in 10,000 live births and leads to mental retardation unless prevented by a special diet.

In a few years, perhaps, genomics will make possible a test of quite different power. The search for human gene variants, and the technology of testing for thousands of gene variants with a single microarray or gene chip, are both progressing so rapidly that experts are already envisaging the possibility of a gene chip capable of scanning a person's genome.[1] Such a genome-wide scan would generate information of enormous significance and complexity, since in essence it would predict all the diseases a person might suffer in the course of his life, with an implicit forecast of life span and likely cause of death. Even for single genes, genetic information could have substantial consequences for both a patient and family members. Genome scans will need to be introduced in the context of substantial genetic counseling and guarantees of confidentiality.

Of the many approaches now being devised to exploit the genome's wealth of information, it is hard to tell which will be successful and when. But the possibilities are so broad that biologists have little hesitation in predicting that the human genome will change medicine.

"I truly feel this is going to revolutionize medicine because we are going to understand not only what causes disease but what prevents disease," says Stephen T. Warren, a medical geneticist at Emory University in Atlanta and editor of *The American Journal of Human Genetics.* "We will understand the mechanism of disease sufficiently to do rational therapy. We will be able to predict who is at higher risk for particular disease and provide advice to individuals as to how best to maintain their health."[2]

Two broad lines of approach are being pursued by genome biologists. One is to scan the normal genome for proteins of medical use. The other is to search for the roots of disease in the variations in the genome. It is differences in the DNA sequence that make each person unique. These differences also underlie a person's susceptibility to disease.

Mining the Genome for Protein Drugs

Most drugs in common use are small chemicals that can be synthesized in chemical plants. These chemicals have the shape and affinity to interact with certain of the body's proteins. But these protein drug targets, laboriously identified over many years by the pharmaceutical industry, amount to fewer than five hundred in number. At one stroke the human genome sequence makes at least thirty thousand proteins available. Although only a few of these will make suitable targets for small molecule drugs, the genome gives the traditional pharmaceutical industry a whole new territory to explore. Since genes often occur in families of similar DNA sequence, many companies are searching through the genome for genes related to those that specify the existing drug targets.

While the traditional pharmaceutical industry seems about to get a new lease on life, a new pharmaceutical industry is emerging that is centered around natural human proteins or protein-based drugs. Typically these are agents such as hormones and cytokines (cell-stimulating factors) that the body itself uses to control behavior. Several proteins had been developed into highly successful drugs well before the age of genomics, such as insulin and the blood-forming protein erythropoietin. The human genome sequence creates the opportunity to find many more.

A number of new companies have sprung up to exploit the medical benefits of the genome. One of the first in the field, with perhaps the most ambitious program, is Human Genome Sciences of Rockville, Maryland. By 2000 the company already had four genome-derived drugs in clinical trials and many more in its pipeline.

Human Genome Sciences was founded by the venture capitalist Wallace Steinberg in 1992 to exploit the shortcut approach to gene finding developed by Venter and known as the EST method after the captured snippets of gene transcripts on which it is based. The weakness of the EST method is that although it efficiently captures transcripts made by the commonly expressed genes in a cell, it may miss those of genes that are turned on only in special circumstances. These very rarely expressed genes, however, are often the most interesting. The genes that build an organism from its egg, for example, are active only at particular stages of development, though some have different roles later in life. Many hormones and signaling molecules are active only briefly, and their gene transcripts can be captured only if the cells have received the right set of prompts from other cells to switch these genes on.

Human Genome Sciences has gone to great effort to capture RNA transcripts from as many types of cells as possible, including fetal cells at various stages of development and body tissues in states of health and disease. The company has also invested in sequencing the full-length genes to which an EST belongs and in some cases in producing the protein specified by the gene. Haseltine has said that the company has discovered evidence for the existence of some 120,000 human genes and has obtained the full-length DNA sequences of 10,000 of them. Although this claim has not been published in the scientific literature for others to assess, an increasing stream of patent applications from Human Genome Sciences gives some substance to the assertion.

Like many other companies, Human Genome Sciences began by searching for novel genes whose DNA structures resembled those of known proteins of pharmaceutical interest. Since many genes occur in families, it was a reasonable bet that if one interleukin gene, for example, existed, there could be many. In fact, Human Genome Sciences found twenty-three new interleukin genes, but so far none has turned out to be a useful drug.

But then the company took a different tack: it decided to focus on a specific broad category of genes, those of the communications network that governs the body's cells and tissues. Though this decision is only just beginning to bear fruit, it could, if successful, have an extraordinary impact on medicine.

For a single celled organism such as a bacterium, each cell is master of its own fate. But in multi-celled animals, the cells must stay in constant communication with one another and must operate under strong self-discipline if the whole organism is to function effectively. The communication system consists of two elements: receptors and signals.

The receptors are protein molecules that are synthesized inside the cell and then inserted into the cell's outer membrane. One half of the receptor sticks out like an antenna, waiting for a specific signaling protein. When the message is received, the inner half of the receptor relays the message to the cell's nucleus where the relevant genes are switched on or off. Most of the signals are also protein molecules.

The system of signals is so important to maintaining the body's health and equilibrium that some 80 percent of present small molecule drugs (meaning simple chemicals, as opposed to proteins such as insulin) work by targeting various components of the system.

Haseltine says that his company has now identified some 11,000 genes, probably the vast majority of those that define the body's communications network, and has determined the complete DNA sequence of most of them.

The capture of the body's entire communications system, if the company's claims are true, was made possible because all of the system's genes possess a common feature. The average human cell possesses about 1 billion protein molecules. To direct each freshly made protein to the correct compartment of the cell after its manufacture by

the ribosomes, nature has arranged for the first few units on a new protein to serve as a signal sequence or cellular zip code that tells the cell's
sorting machinery where to deliver it. Receptors and signaling proteins
share the same zip code because both must be exported from the cell,
the only difference being that the receptors are half exported and then
stuck in the membrane.

The zip code for export is not a simple five-digit number but rather a
sequence of twenty amino acid units that can take a large number of different forms. Still, it is possible to write a computer program that rec-
ognizes the twenty amino acid sequence and all the possible sequences
of DNA bases that correspond to it. By scanning its large collection of
full-length genes, Human Genome Sciences was able to fish out all
genes coding for exportable proteins.

Although this remarkable claim has not been published in the scientific literature, the company has filed for patents on some 7,500 of these
genes. And some experts believe Haseltine's claim is credible.

"Absolutely he is onto something, there is no doubt about that," says
Günter Blobel, the Rockefeller University scientist who won a Nobel
Prize in 1999 for discovering the cell's protein sorting system. "It's obviously of great commercial interest because if you have all the cell surface and secreted proteins you have the major pathways of cell-to-cell
interaction and tissue formation." Blobel noted that the 11,000 proteins
claimed by Human Genome Sciences sounded like the right number,
though there were probably some that the company had missed.[3]

With the body's communications network apparently in hand,
Human Genome Sciences has set about ascertaining the role of each
protein. Many of the proteins made from the 11,000 genes in the system
have been tested for their effects on one hundred different human cell
types. With this database, the company intends to turn drug discovery
into a systematic, rational exercise.

For example, it searched for a protein to make wounds heal, assuming that the body must have some natural set of signals that trigger repair processes when the skin's surface is breached. "We put skin cells in
culture and found fourteen proteins that caused skin cells to grow,"
Haseltine recounted. But many of the body's signaling proteins have a

range of powerful effects. Testing the fourteen proteins on the battery of one hundred different human cell types, the company's scientists found that several had unwanted effects. Some of these proteins had indeed been identified by other companies, which put them into clinical trials only to find they were toxic, Haseltine said. "Only one of the proteins made skin cells grow and did nothing else."[4]

That protein, named keratinocyte growth factor-2, or KGF-2, is being tested in clinical trials for its ability to heal venous ulcers, a type of obdurate skin lesion that affects about 1 million people in the United States and Europe. As of September 2000, the protein was proving safe and effective in a small number of patients and was being advanced to larger scale trials.

KGF-2's healing effect is not confined to skin. It induces growth in all the body's surfaces, the internal lining of the mouth and intestines as well as the skin. Trials are also under way to test its use in mucositis, a serious inflammation of the mucous membranes that affects some 600,000 chemotherapy patients worldwide, and ulcerative colitis, a bowel disease that affects some 300,000 people in the United States.

When skin cells are sitting tightly bound to one another in normal fashion, their growth is naturally inhibited. But when they lose contact with their neighbors as after some wound or trauma, a gene that makes the KGF-2 receptor is triggered, and the wounded skin cells insert the receptor into their membranes. When KGF-2 is detected, the cells start to grow and divide until they again establish contact with one another and the wound is healed.

Since the body already possesses its KGF-2 wound healing process, why does it not make enough KGF-2 to heal any wound, and why should extra doses of the protein help if the body's own dose is for some reason insufficient? The practical answer, if the clinical trials succeed, will be that extra amounts of KGF-2 do in fact help. The theoretical reason, Haseltine suggests, has to do with the evolutionary trade-off the human body has had to make between healing wounds and guarding against cancer.

Letting any cell grow and divide without restraint is a dangerous license because of the risk that it may seize the chance of autonomy and become a tumor, oblivious to its host's wider interests. On the other hand, the more vigorously cells can proliferate, the quicker an animal can recover from a wound. It looks as if evolution balances these interests in a way that depends on the life span of each species. Animals that live only a couple of years before succumbing to cold, hunger or predators do not survive long enough to get cancer, so why not give them a vigorous repair process? But for people, who live many years, a better insurance policy against cancer is merited, which means a much feebler response to wounds. That may be why an extra dose of KGF-2 could be extremely helpful in supplementing the body's own supply.

Because of the power of the body's signaling molecules, their concentration is tightly controlled, and any long term deviation outside the usual range, either above or below it, is likely to cause symptoms of one kind or another. Haseltine hopes to treat both conditions—too much or too little signal—in diseases where an errant signal and its consequences can be linked.

The company believes it has done this with an important signal called BLyS, which it is now testing in various forms for use against three different diseases. BLyS is the long sought signal that makes the cells of the immune system secrete the antibody proteins that attack bacteria and foreign agents. These cells, a class of white blood cells, are called B lymphocytes or B cells; hence BLyS's name, which stands for "B-lymphocyte stimulator."

BLyS is a member of a well known family of cell stimulating factors and on that basis was discovered—and given different names—by five other groups as well as Human Genome Sciences. Companies such as Amgen and ZymoGenetics are also developing medical uses for it.

The first disease with which BLyS may help is lupus erythematosus, a serious autoimmune condition, meaning one in which the immune system starts attacking the body's own tissues. There have long been signs that lupus, and perhaps rheumatoid arthritis, another important autoimmune disease, result from excessive antibody production by the B cells, presumably driven by a surfeit of BLyS.

University researchers working in collaboration with Human

Genome Sciences reported in October 2000 that BLyS is indeed generally elevated in lupus patients, and to some extent in people with
rheumatoid arthritis. The finding lays the basis for a possible treatment:
injections of anti-BLyS antibodies to reduce the amount of circulating
BLyS in the body.

Human Genome Sciences has contracted with the Cambridge Antibody Technology Group of Cambridge, England, to manufacture anti-
BLyS antibodies through a clever laboratory method that mimics the
body's own generation of antibodies against a particular antigen. The
Cambridge company links the human antibody-making genes to a virus
that infects a bacterium. The bacteria then express on their surface a
vast variety of antibody proteins, just as do the cells of the human immune system. From this bacterial library the human antibody is selected that has the desired specificity, in this case for the BLyS protein.
Trials of the anti-BLyS antibody are to begin in 2001. Haseltine says he
expects that half the company's products in future will be antibodies to
the human signaling proteins it has found.

BLyS itself should be useful in another group of diseases in which
the body does not produce enough BLyS and the immune system is im-
paired. The diseases are known collectively as Combined Variable Immune Deficiency.

A third group of diseases in which a role will be sought for BLyS are
the various cancers that develop from B cells at different stages of development. These cancers, known collectively as B-cell lymphomas,
retain many properties of B cells, including, at least in some cases, that
of displaying on their surface the usual receptors for BLyS. When a
BLyS signal docks with a receptor, the B cell pulls both the receptor and
signal inside itself for recycling, behavior that provides an opportunity
to selectively poison B-cell lymphomas. Human Genome Sciences is
preparing BLyS molecules that are linked to a radioactive chemical.
Drawn into the lymphoma cells, the radioactive BLyS molecules are
expected to kill their hosts, a proposition that will be tested in patients
in 2001.

The radioactive BLyS molecules will of course also attack and kill the body's normal B cells. But because only mature B cells display BLyS receptors, the hope is that a patient's immune system will be able to rebuild itself after treatment from the immature B cells.

All of Human Genome Sciences' BLyS proposals are in or about to enter clinical trials, a hurdle many candidate drugs fail. But the range of possible new treatments—all for serious and largely intractable diseases—that have emerged from a single new signaling factor shows how vast a range of possibilities the genome has opened up for imaginative exploitation. Human Genome Sciences may indeed have discovered, as Haseltine is not bashful in claiming, a new paradigm for drug discovery. But, as is evident from the fierce competition in the BLyS field, many other companies seem to be pursuing similar leads.

Haseltine notes that the small molecule drugs on which most large pharmaceutical companies depend are generally foreign to the body and toxic in various degrees. Protein drugs, such as insulin and erythropoietin, are natural products and not inherently toxic to the human body when given in the right doses. On the downside, proteins are generally more expensive to manufacture than chemicals and must usually be taken by injection.

Still, Haseltine predicts that the future belongs to protein-based drugs, with 10 percent of new drugs coming from genomics in ten years and virtually all within thirty years.

Systematic drug discovery is not so stirring a phrase, maybe, but the concept of being able to develop treatments for almost any known disease by looking for the appropriate signal or receptor in the human genome would, if it works, be a remarkable departure from the hit-and-miss approach that has prevailed hitherto. While university biologists are just starting to identify the genes in the human genome sequence, Human Genome Sciences already knows, from its RNA transcript capture experiments, exactly which set of genes is expressed in each type of human cell, Haseltine says. The study of human variation will certainly be important in identifying the gene variants involved in human disease. But that is a long road. In going straight for the gold, Haseltine is confident that Human Genome Sciences will harvest some of the genome's first fruits, while companies that follow Celera's lead in trying to analyze the genome itself are taking the long way round.

Gene Chips and the Anatomy of the Cell

The human cell is a complex miniature machine that performs a large number of sophisticated operations simultaneously. Until the genome sequence was in hand, there was little hope of being able to analyze a whole cell. Most biologists were content to understand a single gene or the set of genes that governed a particular process.

But the genome sequence and devices known as gene chips have together provided a sudden ray of hope for the prospects of understanding the cell's almost impossibly complicated mechanics. "There is every sign that the rapidly evolving technology of the post genome era will unravel the function of the human genome and explain how the 50,000 to 100,000 genes interact with one another and the environment to make us what we are,"[5] writes David J. Weatherall, an expert on the genetics of blood disorders.

A principal use of gene chips is to monitor which genes in a cell are expressed or switched on, so in this application the devices are often called expression chips. An expression chip is a powerful tool because it can show specific changes in the pattern of gene expression when the cell undertakes some operation. By exposing a cell to a hormone, say, and comparing the before and after expression patterns, a researcher can pinpoint the gene or set of genes that is switched on or off by the hormone's action.

The kaleidoscope of gene activity is so complex that expression chips can be hard to interpret. Still, they promise to provide an unrivaled insight into the anatomy of the working cell. And by comparing normal cells with diseased ones, they can show the precise derangement of gene function that underlies the disease, an essential step toward both diagnosing it and understanding its cause.

Gene chips depend on the chemical attraction that any DNA sequence has for its exactly complementary sequence. In other words, two DNA strands will match perfectly, as they do in the double helix of DNA, if every A, T, G, or C base on one strand is matched respectively with a T, A, C, or G base on the other. To prepare a gene chip, DNA strands complementary to those of the genes to be monitored are placed in an array on a glass slide. Captured genes from the cell under test are then washed over the array and will bind tightly to any DNA on the chip whose sequence is exactly complementary to their own.

The binding reactions between the chip-bound DNA strands and the test DNAs can be visualized, usually by tagging the test DNA strands with a fluorescent dye that signals their presence. The chip will then show a pattern of light and dark squares, the light squares signaling that a gene corresponding to each of the sequences in those positions was switched on and the dark squares that the relevant genes were switched off.

In one kind of gene chip, developed by Pat Brown of Stanford University, the DNA strands corresponding to genes of interest are picked out of an organism's genome and amplified. A robotic device then siphons up a sample of each gene and deposits a minute drop on a precisely designated spot gridded out on a glass slide.

The drops of DNA are so tightly packed that, for example, all of the 6,200 known or suspected genes of the yeast cell can be packed onto a single microarray.[6]

Another kind of chip, constructed on a quite different principle, is made by Affymetrix of Santa Clara, California. In the Affymetrix method, DNA chains are built up, base by base, on an array of stalks implanted across a microarray. By use of light-sensitive chemistry and masks similar to those used in the semiconductor industry, the bases added to the stalks in each cycle can be made to differ from one square of the microarray to another. The result is that each of 100,000 or so squares on the finished chip carries DNA chains of a different sequence, as determined by the chip's programmer.

An Affymetrix chip is a square of glass with sides just over half an inch long. The chip is mounted as a window in a tough plastic cassette the size of a matchbook. There are holes for the fluid with test DNAs to be pumped through the chip. The cassette is then mounted in a laser scanner that illuminates each square of the microarray, causing fluorescence in every square where binding has occurred between chip-bound and test DNA. Affymetrix chips are sold for one-time use only, but they are so expensive that much graduate student labor is expended on washing off the test DNAs and preparing the chips for repeat duty.

The matching between the chip-bound DNA and its targets is so precise that if the two differ by a single base they will bind poorly or not at all. Because Affymetrix chips can distinguish every base, they can be used to sequence DNA by comparing a test piece of DNA with one whose sequence has been programmed into the chip. The entire sequence of the human immunodeficiency virus, the cause of AIDS, has been programmed into an Affymetrix chip, as has that of the human gene for BRCA1, a tumor suppressor gene that is often inactivated by mutation in breast cancer. Use of these chips can show the mutations that are important in disease.

Gene chips are still in their infancy. Technologists are squeezing more and more squares into the microarrays so as to increase the number of genes or DNA bases a chip can analyze. Computer programmers are working out algorithms to interpret and seek patterns in the mass of data that chips produce about gene activity. Biologists are learning how to extract reproducible results from their chips since the pattern of gene expression in a cell can vary greatly due to minor causes, such as whether the cells have been kept in the light or dark.

But when all these problems have been sorted out, chips seem likely to become a major tool for interpreting the human genome. Some experts believe it is only a matter of time and improved technique before chips are able to "accommodate the entire population of genes carried within the human genome."[7]

Chips have already proved clinically useful in the diagnosis of cancer. Clinicians have long been puzzled that patients with a cancer of the white blood cells called diffuse large B-cell lymphoma responded in unpredictable ways to chemotherapy, with 40 percent doing well and the others succumbing. A gene expression chip has shown that the lymphoma, of which there are more than 25,000 cases a year, is in fact two diseases, as judged by the pattern of genes expressed in each. Patients with one form did well on chemotherapy, and patients with the other did not. Use of the expression chip should allow physicians to recommend patients in the second group for an immediate bone marrow transplant, the backup treatment when chemotherapy fails.[8]

Probably many other diseases will turn out to be composites of diseases that have different genetic roots but result in the same symptoms. Teasing apart these separate diseases may prove an essential step in finding appropriate treatments for each.

SNPs and the Genetic Basis of Disease

Everyone, so far as is known, has the same set of genes but everyone is different. The reason is that most human genes come with slightly different DNA sequences. These variations can lead to slightly different proteins, with the overall result that no two people are exactly alike save for identical twins. The variations are also the source of people's differing susceptibility to disease.

The human genome sequences that have now been obtained by the public consortium and by Celera provide a point of reference for defining human variation. Just how the human genome varies from one person to another is the subject of active research and a matter of great interest because it bears on human ancestry and population history as well as on the genetic causes of health and disease.

The sequence of nucleotide bases in everyone's genome is pretty much the same. That's no surprise since the ancestral human population seems to have been very small—a mere ten thousand breeding individuals—and to have existed at this size in the very recent past, as evolutionary history goes—perhaps as little as a mere fifty thousand years ago. The ancestral population was not large enough to have much genetic variation within it, and there has not been enough time since for much more variation to have accumulated, despite the enormous expansion of the population.

Genetic change, or mutation, can occur at any of the 3 billion nucleotide sites along the genome. But most such changes are rare and of little importance. Geneticists are particularly interested in the relatively common changes, which are defined arbitrarily as being those that are found in at least 1 percent of the population.

These changeable sites on the DNA are known to geneticists as polymorphisms. Because only a single base is involved, the changes are called single nucleotide polymorphisms, or SNPs, pronounced "snips." SNPs include additions and deletions of a single base, as well as conversions of one base to another. Some causes of mutation, such as radiation and spontaneous chemical change, produce odd chemical groups different from the canonical four bases, but these are immediately spotted by a cell's DNA repair system. The repair system replaces a damaged base with one complementary to the base on the opposing DNA strand, but very occasionally it goofs. The result is a SNP, which, if it occurs in an oocyte or sperm cell, will then be propagated to all the individual's descendants.

Present estimates are that on average about one in every 1,250 sites along the human genome is a SNP, a place where some people (at least 1 percent) may have one nucleotide, some another. As more SNPs are found, the true number in the human population may turn out to be one in every 750 nucleotides or less.

SNPs are the commonest kind of variation in the human genome and the one that seems likely to explain the common diseases, but there are other important kinds of genetic change. Sometimes the DNA copying mechanism gets stuck and erroneously repeats the same few units of DNA over and over. When these repeats occur in the middle of a gene, they can impair the gene's protein. This is the genetic basis of Huntington's disease. Or a whole section of DNA may accidentally be deleted, with the loss of several genes. Williams syndrome is caused by loss of a 1 to 2 million base long sequence of DNA from chromosome 7; because several genes are deleted, patients have a variety of symptoms including, surprisingly, a distinctive character: they are often very friendly, articulate and musical.[9] It is not known why the loss of a few genes should cause such a specific personality change.

The SNPs are spread across the entire genome, although for reasons not yet understood they are more common in some regions than others. Because most of the genome is repetitive DNA of no known function, most SNPs have no effect on the individual who possesses them. Only the SNPs that fall within the protein-coding regions of a gene, or within the control regions of DNA that govern a gene's activity, are likely to make a difference.

Some of these changes may be positive and improve the function of the gene's protein, but these slight enhancements are hard to detect. Most changes are bad, sometimes dramatically so. Sickle cell anemia, for instance, is caused by a single SNP or base change in the beta chain of the oxygen-carrying protein hemoglobin.[10]

Several factors play a role in Alzheimer's disease, but one of them is related to SNPs that occur in the gene for apolipoprotein E, a protein that is found in the bloodstream and is responsible for ferrying cholesterol. The protein occurs in three major forms, known as E2, E3, and E4, of which one, E4, is associated with an increased risk of developing Alzheimer's. The E4 form is produced by a SNP that changes a C to T, with the result that the amino acid unit at position 112 of the protein is changed from a cysteine to an arginine.

Many people who develop Alzheimer's do not have the E4 SNP in their apolipoprotein gene, and many people with the E4 SNP do not get Alzheimer's. In other words, the SNP is neither necessary nor sufficient for the disease. Biologists do not yet understand why it increases the risk of Alzheimer's, but it is an important clue that may lead to a better understanding of the disease's mechanism.

Even when a SNP does not cause disease, as most do not, it may lie in or near a gene variant that does promote disease. Most SNPs of interest are in fact marker SNPs rather than direct disease causing SNPs. It is through these marker SNPs that geneticists hope to identify the variant genes responsible for common diseases.

It is not yet clear how many SNPs will be needed to track down genes of interest. Early estimates of 100,000 have been raised to 300,000. That is the number of SNPs being sought by the SNP Consortium, a group funded by thirteen pharmaceutical companies in the United States and elsewhere together with the Wellcome Trust of London. The SNP Consortium works in collaboration with the National Institutes of Health and, like the public consortium that sequenced the human genome, makes all its data freely available.

Meanwhile, continuing the human genome race on another front, the public consortium's rival is also pursuing SNPs. As of September 2000, Celera Genomics was offering its clients 2.8 million human SNPs, of which it had found 2.4 million itself and had downloaded 400,000 from the public databases. By January 2001, Venter said, his company possessed 4 million SNPs.

A large number of SNPs is needed to locate disease-causing variant genes, in part because of the sheer size of the genome but also because the genome gets shuffled in each generation. The paternal and maternal chromosomes exchange equivalent sections of a chromosome in a more or less random fashion. Suppose a mutation occurs in some individual that changes a gene's protein in a damaging way. In each of the individual's descendants the marker SNPs that lie near the mutated gene may be exchanged for a section of DNA that does not carry such a SNP. The farther away the marker SNP lies from the gene, the greater the chance it will be separated from the mutation with the passage of generations. So to have a reasonable chance of finding a marker SNP still linked to the ancestral gene mutation, a geneticist needs to have as many SNPs as possible. Another complication is that a particular gene may be rendered defective by any one of many different mutations, each of which has occurred in a different ancestor and with a different pattern of neighboring marker SNPs.

Some disease-causing gene variants are relatively easy to find, but the diseases they cause tend to be rare. The variants have such a strong effect that they alone are sufficient to cause the disease. These single-gene maladies are called Mendelian diseases because they are inherited in simple patterns like those seen by the abbé Mendel in his peas.

About a hundred Mendelian gene variants have now been identified.

But these Mendelian diseases, though serious enough for those who have them, are not a major burden of disease on the population as a whole. They are so serious, in fact, that they tend to be lost quickly from the population and arise again as new mutations. The mutations that cause many of these Mendelian diseases are so rare that they don't even count as SNPs since they appear in less than 1 percent of the population.

The common diseases, such as most cases of heart disease, cancer, diabetes and psychiatric disorders, are thought to be caused by several genes acting in concert. Each of the genes, however, generally makes only a small contribution to the overall disease, and the slightness of the effect makes the gene very hard to find.

Only recently has the first gene affecting a multigenic disease, type 2 diabetes, been identified by scanning the genome. Other such genes, notably the apolipoprotein E4 gene variant that is involved in Alzheimer's disease, have been detected by examining genes whose known roles made them likely candidates in the disease. But the candidate gene approach has limitations because many diseases are caused by biochemical pathways that are unknown and the genes that control them have not been identified.

The hunt for the type 2 diabetes gene illustrates some of the complexity of human disease genetics, as well as the power of the new tools to dissect it. Type 1 diabetes arises when the body's immune system mistakenly attacks the pancreas gland cells that produce insulin, the hormone that induces cells to take up glucose from the bloodstream. Type 2, the commoner kind, which affects almost 6 percent of the American population, is really a large group of different diseases with a common end point, that of a failure to metabolize glucose correctly.

Each of these component diseases is presumably caused by a different variant gene or set of variant genes. Some of these gene variants are in fact of the Mendelian type; a single gene variant has such a strong effect that it alone is enough to derange sugar metabolism. One of these genes makes a protein enzyme called glucokinase, which helps the insulin-making cells of the pancreas detect how much glucose is in the bloodstream and secrete insulin accordingly. Geneticists have catalogued a large number of different mutations that can occur in the glucokinase gene, any one of which can cause type 2 diabetes. There are four other proteins that, when dysfunctional, can alone cause the disease; all are thought to be involved in the proper development of the pancreas gland.

Like most Mendelian disease genes, these single-gene causes of type 2 diabetes are quite rare and together account for only 5 percent of cases. The rest, the vast majority, are thought to be caused by several different gene variants acting in concert. The first successful search for one of the gene variants involved in the multigenic version of type 2 diabetes was reported in October 2000 by biologists at the University of Chicago and the University of Texas Health Science Center in Houston.

Craig L. Hanis at Texas had been collecting diabetes related samples for nineteen years from Mexican-American families in Starr County, Texas, among whom there is a high incidence of the disease. In 1992 he joined forces with Graeme I. Bell of the University of Chicago, a geneticist who specialized in diabetes. By comparing the genomes of matched pairs of Mexican-American siblings, one of whom had diabetes and the other didn't, they implicated a region of DNA at the end of the long arm of chromosome 2 as carrying a causative gene variant.

This region of DNA was nonetheless a vast territory, 1.7 million bases in length, and it contained 7 known genes and 15 suspected ones. No one analyzing a multigenic disease had been able to take the next step, that of identifying a specific gene within the general chromosome region indicated by genetic studies.

There are many sites of DNA variation along the 1.7 million base sequence, and a colleague of Bell, Nancy J. Cox, developed a statistical way of measuring how strongly each was linked to the appearance of diabetes in the patients. She narrowed down the search to a previously unknown gene. The gene was named calpain-10 because its DNA sequence resembles a family of known genes that specify proteins called calpains.

The intricacy of the human genetic machinery is shown by the fact that the three single nucleotide variations associated with diabetes each occurred in an intron of the calpain-10 gene. Introns are the alternating sections of a gene that are spliced out from the exons, leaving only the joined up exons as the message that goes to the protein-making machinery. The introns can thus play no role in specifying the calpain-10 protein. However, they may be involved in some unknown way in controlling the activity of the gene.[11]

Calpain-10 is a novel gene and no one yet knows its role in glucose metabolism or how the variations found in the Mexican-American patients derange the gene so as to contribute to diabetes. Statistical tests suggest that the calpain-10 variations account for 14 percent of the diabetes risk in Mexican Americans and 4 percent of the diabetes risk in Europeans. It is not surprising for different DNA variations and different risk factors to occur from one ethnic group to another.

Discovery of the calpain-10 gene's role in diabetes underlines both the promise and the difficulty of identifying disease-causing gene variants through the human genome. Most of the Hanis-Bell team's eight-year gene hunt was undertaken before the human genome sequence or SNP collections became available but would probably have proceeded considerably faster if the genome had been available.

"If they were starting this today, it would be profoundly different. The human genome sequence would be in hand. There are a million SNPs they could just pull off the internet. This whole long drawn out process could be telescoped into a few months, and one hopes that other hunts can find their target more quickly," says Francis Collins of the National Institutes of Health.[12]

Finding a genetic variation at the root of diabetes is merely the first step in developing a cure. The calpain-10 gene presumably controls some unexpected aspect of glucose metabolism. Biologists must now work down the chain of events from the variations in the gene's third intron to its role in diabetes, hoping that some link will prove vulnerable to a drug or other intervention.

The Icelandic Window

One of the most interesting searches for disease genes is a wonderfully inventive experiment designed by Kári Stefánsson. An Icelandic neurologist, Stefánsson conceived the idea that the special properties of Iceland's people and history made the population particularly suitable for tracking down the genetic roots of complex diseases.

Icelanders are descended from a small number of founders, probably 8,000 to 20,000 people, who discovered and settled the islands between A.D. 870 and 930. Some 75 percent of the men were probably Vikings from Scandinavia and the Viking settlements in Britain.[13] Most of the women were Gaelic, brought as wives and slaves from Britain. The women seem also to have included people from many other parts of Europe, so the founding population was probably not particularly homogeneous,[14] despite Stefánsson's initial belief that Iceland possessed "unique genetic homogeneity." Still, the genetic diversity was probably substantially reduced by three disasters that befell the population: an outbreak of pneumonic plague in 1402–4 that carried off 45 percent of Iceland's inhabitants; a 35 percent reduction in 1708 caused by smallpox; and a 20 percent reduction in 1784–85 because of the famine that followed a volcanic eruption.

Because few if any new settlers arrived after the founding in the tenth century, most Icelanders can trace their lineages back to a founding ancestor. The country has extensive health care records of every major illness going back to 1915 and centralized autopsy samples that date back for half a century. All these features make the Icelanders excellent subjects on which to try to trace the pedigrees of complex diseases. The reason is that disease-causing gene variants are much easier to identify if everyone in a population inherited the variant from the same ancestor. There is a good chance that they also inherited stretches of DNA on either side of the gene along with their markers—the geneticists' word for any recognizable piece of DNA such as a SNP. If apparently unrelated people with schizophrenia, say, have a particular pattern of SNPs, the SNPs may well lie close to a gene variant that contributes to the disease. In larger populations, the particular gene variant may be surrounded by a variety of different SNPs, making it much harder to locate.

Stefánsson formed deCODE Genetics in 1996 as a "population-based genome company," with the idea of running a national health database for the Icelandic government and at the same time looking for disease related genes. In return, the government granted the company exclusive rights to market the database abroad for twelve years.[15] The idea was of such interest to the Swiss pharmaceutical firm Hoffmann–La Roche that in 1998 it paid more than $200 million for the rights to any genes deCODE might discover to be involved in such diseases as schizophrenia, heart disease, Alzheimer's, and emphysema.

The scheme drew criticism both in Iceland and abroad, chiefly on the ground that it would be impossible to protect the privacy of the Icelanders whose health records were being mined. In December 1998, after a furious debate, the Althingi, the Icelandic parliament, approved a revised bill to create the health care database. Despite the controversy, Icelanders seemed to believe that the benefits of the deal—new jobs and revitalized medical research—outweighed the risks. By June the following year, only 9,000 of the island's 270,000 inhabitants had exercised their right to opt out of the database.[16]

DeCODE Genetics subsequently reported considerable success in locating disease-causing gene variants for Roche, although the claims are hard to assess because articles describing the claims have not yet appeared in the scientific literature. According to its Web site,[17] by the end of 2000 the company had identified gene variants for nine diseases, including Alzheimer's, multiple sclerosis, schizophrenia, osteoporosis, and osteoarthritis. Roche was pleased enough with deCODE's progress that it signed up for a second venture in March 2001, a five year collaboration focusing on diabetes, cancer and autoimmune diseases.[18]

The schizophrenia-related gene was identified by comparing four hundred Icelandic patients with the same number of unaffected family members. Roche said it hoped to use the information to develop new ways of diagnosing and treating the disease, which affects more than 0.5 percent of people worldwide.

DeCODE's success has also been noticed in Estonia. The Estonian Genome Foundation hopes to compile DNA data and health profiles of the Baltic state's 1.4 million citizens and raise $100 million by selling access to the data. Since each ethnic group probably has its own slightly different pattern of disease proclivities, plans to genotype populations may move forward in other countries if Iceland's example is successful and if satisfactory ways to ensure confidentiality and informed consent can be worked out.

The Promise of Pharmacogenomics

While biologists seek the SNPs that underlie disease, pharmaceutical companies are hoping to use SNPs to match patients with drugs.

Medicine is sometimes said to be as much an art as a science, and one reason for the artistic element is the unpredictable way in which patients respond to medicine. Some get better, some are not helped much by a drug. And many patients suffer from a variety of drug side effects, some extremely serious. According to a 1998 survey, 106,000 patients die every year and 2.2 million are injured because of adverse reactions to prescribed drugs.[19]

Pharmacogeneticists have long understood part of the reason for the variability of patient response, without being able to do much about it. In the liver's cells are a group of enzymes produced by a family of important genes known as cytochrome P450s. Each member of the family specializes in metabolizing, or chemically degrading, a particular category of chemicals, and together the family of cytochrome P450s breaks down and eliminates from the bloodstream most of the drugs in current medical use. One member of the family, which labors under the obscure name of cytochrome P450 CYP2D6, is alone responsible for metabolizing about 20 percent of all commonly prescribed drugs, including many of those given for psychiatric and cardiovascular disease.

Despite the importance of these genes, many people possess mutated forms of them, meaning that there are SNPs in the genes, some of which render the gene's protein less effective. When the CYP2D6 gene is inactive, people break down the drugs it metabolizes very slowly and so are exposed to much higher and maybe toxic doses. Some people have a genetic change with the opposite effect: they carry several extra copies of the whole gene. The result is that the drugs are cleared from the bloodstream so rapidly that it is hard to maintain a therapeutic dose.

One explanation offered for the common mutations in the cytochrome P450 genes is that the genes' original purpose was to get rid of the many plant toxins in the early human diet. Now that a much less toxic group of plants is consumed in the modern diet, inactivating mutations that arise in these once essential genes are not being eliminated and the genes are gradually disappearing in the usual manner of genes for which evolution has no further use.

Drugs also come into contact with many other types of proteins, which may also vary from one person to another. The variations in the cytochrome P450s are the best known because they have been studied longest.

The promise of pharmacogenomics now becomes clear: in principle the technique could maximize the benefit of every drug and eliminate its side effects by genetically screening the population and matching drugs to patients in the best possible way.

"The medical significance and economic value of a simple, predictive medicine response profile, which will provide the information on the likelihood of efficacy and safety of a drug for an individual patient, will change the practice and economics of medicine," says Allen D. Roses, director of genetics at the Glaxo Wellcome center in Research Triangle Park, North Carolina.[20]

By genotyping patients—screening their genomes for important SNPs and other variations—drug companies may be able to avoid the adverse side effects that rule out many promising drugs in clinical trials. Drugs that come with an obligatory genotyping test will be more expensive. But if outcomes are improved and side effects reduced, the extra cost may be justified.

Should genotyping become widespread, drug companies may find themselves in a strange new world of designing and marketing drugs to many different subgroups of the population, instead of the one-drug-fits-all approach that has prevailed until now. Curiously, many of the mutations in the cytochrome P450 family of genes vary widely in frequency from one ethnic group to another, perhaps reflecting the random process in which a once necessary gene is phased out. The CYP2D6 gene is inactive in about 1 percent of Japanese, 3 to 10 percent of Caucasians, and 15 percent of Nigerians.[21] "In Britain several million people are thus at risk of compromised metabolism or adverse drug reactions when prescribed drugs that are CYP2D6 substrates," three British pharmacologists noted in a recent article.[22] Only 1 percent of Caucasians have extra copies of the gene, but 29 percent of Ethiopians do.

"All pharmacogenetic polymorphisms studied to date differ in frequency among ethnic and racial groups," two pharmacologists wrote recently in *Science,* noting that these differences "dictate that race be considered" in studies aimed at associating particular genotypes with disease risk or drug toxicity. They predict that over the next decade pharmacogenomic advances made possible by the human genome sequence should "permit the development of therapeutic agents targeted for specific, but genetically identifiable, subgroups of the population."[23]

A taste of how powerful pharmacogenomics may prove to be, as well as of the complexity of human genetics, emerges from a recent study of how asthma patients respond to albuterol. Albuterol is a widely used drug that is inhaled to make the muscles of the airways relax during an asthmatic attack. It works by acting on a receptor protein that straddles the wall of the bronchial muscle cells.

Patients respond in very different degrees to albuterol. For some it works very well, for others hardly at all. The difference lies to a large extent in the receptor protein, which is known as the beta-2 adrenergic receptor, and in genetic variations in the gene that makes it.

Stephen B. Liggett of the University of Cincinnati sequenced the gene in normal and asthmatic individuals drawn from four major ethnic groups and found thirteen sites at which some people had one base, some another. These sites of DNA variation, or SNPs, could in principle occur in a large number of combinations but Liggett found only twelve combinations in the people he studied. Geneticists refer to any combination of SNPs as a haplotype. African Americans mostly had the haplotypes that Liggett numbered 1, 2, 4, 6, 9, and 12. Asians had 1, 2, 4, 6, and 11. Haplotypes 1, 2, 4, 6, 7, 9, and 12 predominated among Hispanic Latinos; and Caucasians had 2, 4, and 6.

Since everyone receives a full set of chromosomes from each parent, there are two copies of each gene, and generally both are active in a cell. The two genes can have the same or different haplotypes. Liggett and his team found that almost all—88 percent—of their asthmatic patients had a particular pair of haplotypes in their two receptor genes. These were the haplotype pairs 2/4, 2/2, 2/6, 4/4, and 4/6. Among these pairs, there was a clear pattern of response to albuterol, with the 2/2 patients responding well, the 2/4 and 2/6 less well, and the 4/4 worst of all.[24]

Liggett's study seems to be one of the first to show how directly the human genome and DNA sequencing can help predict patients' responses to drugs. Interestingly, none of the SNPs in the receptor gene had any predictive power by itself; it was only when the SNPs were considered in the combinations—haplotypes—in which they occur in the human population that the correlation with drug response emerged.

By measuring the response of people with the various haplotype pairs to other asthma drugs, Liggett and his colleagues hope to develop a precise method of deciding which drug is best for each patient. "This is a step toward personalized medicine. The era of one drug fitting all is probably coming to an end soon," Liggett said.[25]

The Genome and Cancer

Cancer is a large cluster of diseases, of which at least one hundred distinct types have been recognized. All result from cells that have accumulated a set of genetic changes that allow them to evade the body's usual strict constraints on growth and proliferation.

Biologists have made considerable progress over the last two decades in understanding the nature of these genetic changes. Based on these deep insights, a few novel treatments are slowly beginning to emerge. Availability of the human genome sequence is likely to give enormous impetus to these efforts.

Among the genome's first applications in cancer may be the use of gene chips to monitor the activity of genes in tumor cells. Gene chips, as noted above, can track tens of thousands of genes at the same time, providing a previously unattainable insight into the overall behavior of the cell. Biologists hope that gene chips will pinpoint the specific genes that are the causes of the various types of cancer. Once such genes are identified, their protein products will become targets against which chemists can try to develop drugs.

Gene chips will also help tremendously in diagnosing the many different types of cancer. At present this is often done either by looking at cancer cells under a microscope or by analyzing marker proteins, such as the PSA antigen of prostate cancer. But visual appearance goes only so far, and the marker proteins are the products of a single gene and do not reflect the full complexity of cancer cells in which a whole assortment of genes is deranged.

As biologists gain skill in analyzing the patterns of gene expression revealed by gene chips, it should eventually be possible to diagnose each patient's cancer with precision, identifying the principal set of gene changes that underlie the cancer and devising a tailor-made remedy to suppress and even eradicate the errant cells.

The present treatments for cancer are far from perfect. Surgery often fails to remove all cancerous cells from the body. Chemotherapy and radiation attack all dividing cells, whether cancerous or not, and damage the body's healthy as well as cancerous tissues. As a genetic disease, cancer almost certainly needs to be attacked on the genetic level, through an understanding of the genetic damage in each type of tumor.

Full fledged cancer cells differ from normal cells in complex ways because their genomes are often altered extensively. Their chromosomes may be rearranged, with regions in which a batch of genes has been amplified many times over. In this state of genomic instability, all the normal cell's usual controls have been overridden or subverted. The cancer cell can grow and divide independently, signal the body's tissues to build a blood supply to its growing mass, burst out of the normal boundaries of its tissue, and set cancerous cells free to colonize other sites in the body.

Biologists believe that most cancer cells follow a definable path to this state of genomic instability. There are at least six steps in cancer, in each of which the cell acquires a genetic mutation that overrides one of the cell's normal constraints. A person may inherit a cancer-favoring mutation, but it need not be realized unless during their lifetime he or she acquires in some cell another mutation, whether by a DNA copying error or by radiation or chemical change. A cell that has acquired two mutations, knocking out two control genes, may then start dividing slightly faster. Any of its progeny that develops a third mutation will gain a growth advantage over the others, and natural selection will continue to favor cells that accumulate the other mutations necessary for cancer to occur. The process of acquiring the six mutations may take many years, which is one reason why cancer is most common later in life.

One of the six changes is for cells to become independent of external growth signals. Usually cells divide only when they receive a specific signal to proliferate from other cells. Many cancer cells have acquired a mutation in the gene for the receptor protein that receives a growth signal, or in one of the intermediary proteins that conveys the growth message from the receptor to the genome in the cell's nucleus. Either kind of mutation can derange the cellular mechanisms that turn a message off after delivery, with the result that the growth message is left on permanently and encourages excessive proliferation.

Over the course of evolution cells have developed a sophisticated network of interacting proteins that serves to detect any abnormality in the cell's division process, such as unusual growth signals, invasion by a virus, or damaged chromosomes. At the center of this network are two complex proteins known as p53 and Rb. Most, perhaps all, cancer cells must disable their p53, either by acquiring a mutated copy of one of the cell's two p53 genes or by indirect means.

One of the body's last ditch defenses against cancer is a mechanism that counts the number of times a cell has divided and forces it into senescence after it has used up its quota. Any incipient tumor that has grown above microscopic size will hit the wall imposed by the division counter mechanism, unless it has managed to switch on a special enzyme called telomerase that can override it.

Another ability the tumor must then acquire is that of switching on a signal that stimulates the growth of blood vessels to deliver nutrients. And then the cells must learn to burst beyond their tissue boundary and colonize sites elsewhere in the body.

Some cancer experts believe that most types of cancer must acquire each of these six hallmarks of cancer cells, although they may do so in different order. In ten to twenty years' time, say Robert Weinberg of the Whitehead Institute and Douglas Hanahan of the University of California, San Francisco, it will be routine to prepare genomewide gene expression profiles of tumors to see exactly which genes have been subverted in order to acquire these six hallmark properties. After this precise diagnosis of an individual tumor, the appropriate drugs could be applied. "We envision anti-cancer drugs targeted to each of the hallmark capabilities of cancer," the two biologists say.[26]

Justification for such optimism can be seen in two new anticancer drugs that have been developed on the basis of precise genetic understanding of cancer cells. Both are targeted against proteins involved in the first of the steps described above, that of the cancerous cell acquiring independence from reliance on growth signals from the outside.

The two drugs are Herceptin, used for a certain category of breast cancers, and Gleevec, for chronic myelogenous leukemia.

In about a third of breast cancer cases, the gene for a receptor protein called epidermal growth factor 2 is amplified and produces an excess of receptor protein. The receptor proteins, embedded in the cells' walls, keep signaling them to proliferate. Herceptin, made by Genentech of South San Francisco, is a pure form of antibody, called a monoclonal antibody, which attacks the segment of the receptor protein that sticks out of the cells and stops them from multiplying.

In chronic myelogenous leukemia, one of the four most common types of leukemia, two chromosomes, numbers 9 and 22, swap sections. The point of interchange occurs within a gene. The gene's product is an intermediary protein in the signaling system for conveying messages received from other cells. After the swap, the gene produces an abnormal product that cannot be shut down as usual after delivering its message. The product, known as a tyrosine kinase, serves as a permanently switched on signal that upsets the cell's equilibrium in various ways, one of which is by causing persistent growth and proliferation.

Researchers at Ciba-Geigy, now Novartis, started in 1990 to look for chemicals that would inhibit the abnormal kinase. They randomly screened a large number of chemicals and then chemically modified the most promising candidates. Six years later, they had developed a class of drugs that blocked the abnormal kinase's action but did not interfere with the many other kinases of the cell's internal communications system.

One of these drugs, called STI-571 and later Gleevec, had dramatic results in early trials, causing 96 percent of chronic myelogenous leukemia patients to enter remission within a month of treatment and with minimal side effects after five months.

The drug also targets a quite different signaling protein, named c-kit, which goes awry in a rare and deadly stomach cancer called gastrointestinal stromal tumor, or GIST. In one trial, some 60 percent of patients with GIST went into speedy remission after taking the drug. It seems that the cancerous cells in both chronic myelogenous leukemia and GIST become very dependent on their abnormal signaling proteins, and self-destruct when the signaling proteins are jammed. Gleevec is also active against a third kind of signaling protein, known as PDGFR-beta, which is deranged in glioma, a kind of brain cancer. Trials of Gleevec against glioma are now in progress.

Gleevec was approved by the Food and Drug Administration in May 2001 after an extremely quick review.[27]

Both Herceptin and Gleevec were developed by conventional methods of molecular biology and drug development, before the sequencing of the human genome. But the genome is a force amplifier that should enable biologists to find any other suitable targets for the drug by scanning the genome, a task that could have taken years without having the sequence at hand. "The emerging field of disease genomics will likely identify other members of the protein kinase family as candidate disease genes," comment Brian J. Druker and Nicholas B. Lydon, two researchers involved in developing the drug.[28]

Devising special methods to attack each of a hundred different types of cancer will be a lengthy undertaking, but there is always the hope of finding some generic flaw common to all or many cancers. Solid tumors need to send out signals to develop their own blood supply, a process called angiogenesis. Many companies and academic researchers are trying to develop drugs that inhibit angiogenesis. Researchers at Genentech discovered the natural signal sent out by many cells, including tumor cells, that need more oxygen. The substance is called vascular epithelial growth factor, or VEGF. Genentech has prepared an anti-VEGF monoclonal antibody that has shown promise in clinical trials of metastatic colorectal cancer.

Besides the need to build a blood supply, another common feature of many tumor cells is the deactivation of the cell's stress detection network. The remarkable protein at the center of this network, p53, is usually quiescent, but it springs into action when the genome is damaged by agents such as radiation, when incorrect growth signals are detected in a cell, or when a cell is affected by chemotherapy drugs or ultraviolet light.

Some biologists believe that nearly all tumor cells must deactivate p53 either directly or indirectly. Once p53 has attained high levels in a normal cell, it can switch on several sets of genes, any one of which will halt an incipient tumor cell in its tracks. One set of genes blocks the cell cycle, the round of events by which a cell divides into two daughter cells. Another gene blocks genes whose products recruit new blood vessels. And p53 can also trigger the self-destruct mechanism that is carried by every cell and that can reduce it to fragments in as little as thirty minutes.

This elaborate network of interactions works by means of chemical groups that are added to or removed from the p53 protein, affecting its state of operation. As three experts on the biology of p53 recently pointed out, none of this can be predicted from analysis of the genome. "Even a complete characterization of the genome (all the genes in an organism), the transcriptome (the genes that are actually expressed . . . at a given time) and the proteome (the proteins that are produced from the expressed genes) would not provide a very accurate portrait of the state of the p53 protein in any cell. The condition of this protein cannot be accurately predicted from just its sequence as it is extensively 'decorated' by different chemical groups, rather as a Christmas tree is decorated by lights and tinsel," the biologists said.[29]

The genome sequence will certainly not by itself answer every biological question. But it provides an invaluable resource for biologists trying to unravel the complexity of the cell circuitry, such as the p53 network. By helping to identify all the genes that may be involved in a process, the genome turns a potentially insoluble morass into a finite problem.

When a cancer-promoting gene is turned on, drugs like Gleevec can be developed to inhibit it. But when a cancer-inhibiting gene like p53 is turned off by mutation or other damage, no drug can restore the gene's tumor-suppressing properties. A quite different approach—using mutated p53 to identify and destroy the cancerous cell—has been developed by Frank McCormick, director of the cancer center at the University of California, San Francisco, and co-founder of Onyx Pharmaceuticals. Drawing on two decades of research on the biology of viruses and cancer cells, McCormick saw an opportunity in the fact that both cancer cells and adenoviruses, one of the causes of the common cold, must inactivate the p53 network. To get itself replicated, an adenovirus must trick the cell it has infected into dividing. Since p53 would normally prevent cell division on detecting the virus's presence, the adenovirus possesses a gene whose product disables p53.

Imagine an adenovirus whose p53-sabotaging gene has been disabled. Such a virus couldn't grow in normal human cells because the p53 network would detect it and force the cell into suicide. But the disabled virus could grow in cancer cells, because they have already disabled the p53 network. So could a disabled adenovirus be the long sought discriminating weapon that would kill tumor cells and spare normal cells?

Onyx-015 is the commercial name of the disabled adenovirus. It has shown remarkable success in preliminary trials for treating advanced head and neck cancer and is also being tested against ovarian cancer, pancreatic cancer, and metastatic liver cancer.[30]

Onyx-015 may fall at the final hurdle, as many drugs do, but should it succeed it will be an entirely novel kind of cancer therapy and among the earliest rewards of biologists' growing understanding of the cancer cell.

The availability of the human genome should greatly accelerate the search for the genetic roots of cancer. By using gene chips to follow the changes in gene expression as cancer cells become metastatic, or able to spread to other sites in the body, biologists at the Massachusetts Institute of Technology discovered a specific gene that may play a role in this gain of function.[31]

The MIT biologists then applied their gene chip technique to analyzing the pattern of genes expressed in acute leukemias and showed that the chip could distinguish between acute myelogenous leukemia and acute lymphoblastic leukemia.[32]

The use of chips to diagnose cancer has been taken to a practical stage by another team of biologists who, as mentioned above, have shown how to distinguish between two forms of large B-cell lymphoma. One form of the disease responds well to the standard treatment and the other does not, an important clarification of considerable use to the physician.[33]

Much research lies ahead in developing standard gene expression profiles of each type of cancer. But precise profiling of cancer with gene chips now seems an attainable goal and one that is likely to bring significant advances in diagnosis and treatment of a long-intractable disease. "We can see great promise in molecular profiling," write two National Cancer Institute biologists, ". . . in the fields of cancer diagnostics and patient-tailored therapeutics. We hope that this technology is not far away from routine use in the clinic."[34]

The initiatives already under way show the rich variety of opportunities the genome offers for understanding disease and improving human health. The human genome sequence won't enable biologists to cure every disease overnight. Each disease presents its own set of problems and requires its own special solution. But the human genome sequence provides, for the first time, a systematic basis for tackling every affliction that the flesh is heir to.

5
Regenerative Medicine

Consider a new way of healing the body:

❖ Instead of cutting the flesh with scalpels, poisoning it with chemical drugs, or burning it with radiation, the physician would gently treat the body with nothing but cells and proteins, seeking to mend like with like.

❖ Instead of depending only on his own knowledge, the physician would seek to tap the information in the genome and to exploit the fact that the body's cells are designed to be a self-assembling system when given the right signals.

❖ Instead of sending a patient home merely patched up enough to live with disease, the physician would not rest until the damaged tissues were replaced with ones as good as new, if not better.

Some biologists believe such a therapy is so close to reality that they have given it a name: regenerative medicine. Regenerative medicine would draw on the rapidly expanding knowledge about a special class of human cells, known as stem cells, and on genome-derived signals that influence cell behavior.

"When we know, in effect, what our cells know, health care will be revolutionized, giving birth to regenerative medicine—ultimately including the prolongation of life by regenerating our aging bodies with younger cells," says William Haseltine, chief executive of Human Genome Sciences.

Thomas Okarma, president and CEO of the Geron Corporation, sees regenerative medicine as a "truly new therapeutic paradigm" whose effects he likens to having a car come back from the auto shop with its malfunctioning parts wholly replaced with new ones.[1]

Ronald McKay, a stem cell expert at the National Institutes of Health, believes the use of stem cells to regenerate diseased tissues and even whole organs is imminent. "In a few months it will be clear that stem cells will regenerate tissues," McKay said in November 2000. "In two years, people will routinely be reconstituting liver, regenerating heart, routinely building pancreatic islets, routinely putting cells into brain that get incorporated into the normal circuitry. They will routinely be rebuilding all tissues."[2]

McKay's confidence is based in large part on the fact that stem cells are very sensitive to signals they receive from surrounding cells and will often develop into the kind of cell the body requires at that position. "I don't know how to make a heart," he says. "But once you know how to take stem cells and turn them into heart muscle, it's easy."

These hopes are not as far-fetched as they may sound. Stem cells from bone marrow have been used for many years to reconstitute the blood system of patients whose own blood-forming cells have been destroyed by chemotherapy or radiation treatments for cancer. And researchers have already laid considerable groundwork in animal experiments for treating Parkinson's disease patients with stem cells that would replace the lost dopamine-producing cells of the brain.

Stem cells differ from the body's ordinary cells in that they are not committed to a specific fate. Fate is a strange term for biologists to use; the idea comes from the fact that human cells, as the body develops, are selected for particular destinies that, once chosen, can never be reversed. Once a cell becomes a heart muscle cell or a cartilage cell or a neuron, it is committed to that fate until it dies.

Stems cells obey quite different rules. Their role is being worked out in a burst of new discoveries, many of them made in the last few years. Only in 1998 did biologists first learn how to grow human embryonic stem cells in the laboratory. A different kind of stem cell, known as adult stem cells, occurs in many tissues of the body. Biologists suspect that each organ and tissue may have its own reservoir of stem cells. But their hiding places are still elusive. Although the bone marrow has long been known to harbor the stem cells of the blood system, the site of the brain's stem cells was discovered only in 1999 and the skin's stem cells in 2000.

There are several different approaches for putting stem cells to medical use, involving both embryonic stem cells and adult stem cells. Like all the body's other cells, adult stem cells must be descended in some way from the embryonic stem cells, but the exact connection between the two types of stem cell has not yet been worked out.

Embryonic Stem Cells

The fertilized human egg is a single cell, and its first act is to divide into two cells, then four and eight. The cells at the eight-cell stage are all equivalent and are termed totipotent, meaning all-powerful, because they can form any and all of the body's different cell types. One of the cells can be removed at this stage for genetic diagnosis, and the remaining seven will go on to form an infant as if nothing had happened.

By the fourth day and the seventh round of division, there are sixty-four cells that begin to transform themselves from a blob into a definable structure. Most of the cells organize themselves into a hollow sphere. The rest, from fifteen to twenty of them, form a clump within the cell called the inner cell mass. The cells now have different fates. Those in the hollow sphere will form the placenta; the cells of the inner cell mass will generate all the tissues of the infant to be. As the twenty or so cells of the inner cell mass grow and divide into the 100 trillion cells of the human body, they transform into many different types of cells—at least 260 have been identified—that are specialized to form different tasks depending on which organ or tissue they are assigned to.

The embryo at this stage is called a blastocyst and has not yet implanted in the wall of the uterus, where it will gestate. When cells are taken from the inner cell mass and grown in the laboratory, they are called embryonic stem cells. Research with the cells presents both technical and ethical problems.

The blastocyst, which has the potential to grow into a person, is destroyed when its inner cell mass is removed. Many in vitro fertility clinics possess freezers full of abandoned blastocysts that have no prospect of coming to term. The clinic mixes sperm and eggs in glassware to generate blastocysts for implantation in the womb but often produces more than are immediately needed in case the first implant doesn't take. After several years the clinic may wish to clear space in its freezers and the couple who owns the blastocysts may indicate they have no further need for them.

Many biologists believe that the medical benefits likely to spring from embryonic stem cells justify the destruction of a few abandoned blastocysts, provided that the parents or owners have freely assented to the procedure. Because the cells grow and multiply indefinitely when cultured in the laboratory, very few blastocysts need be destroyed. In principle, just one blastocyst could supply all the embryonic stem cells that researchers would need. In practice, researchers would like to have several different cell lines available, just to make sure they were not dealing with some unusual type. Perhaps as few as a hundred or so would suffice, meaning only that number of abandoned blastocysts would be destroyed. Many opponents of abortion, however, believe that any destruction of a potential human life is impermissible.

Research with human embryonic stem cells was for a long time impossible because the cells could not be kept stable in the laboratory. They would spontaneously differentiate, meaning that they developed into mature cells of different kinds. Adapting a method developed for the mouse, James Thomson at the University of Wisconsin reported in November 1998 that he had successfully grown human embryonic stem cells. A key feature of the technique is blocking the spontaneous differentiation of the cells with a special inhibitory factor.

At the same time John Gearhart of Johns Hopkins University, using a similar method, cultivated human embryonic germ cells. These came not from blastocysts but from human fetuses that had been aborted for therapeutic reasons. Before the cells of the inner cell mass start to develop, a small pocket of cells is set aside to generate the germ cells, whether egg or sperm. These embryonic germ cells migrate to a special niche in the fetus, where they are protected from further development. It was these cells that Gearhart isolated. The embryonic germ cells are very similar to human embryonic stem cells. Their ethical status is less problematic, since they come from pregnancies terminated for medical reasons, but they are much harder to obtain than the cells derived from blastocysts.

Embryonic stem cells are of enormous importance because they can form probably all of the tissues of the body. When the special inhibitory factor is removed from their culture medium, they immediately start to differentiate into many different lineages, leading to cells such as neurons, skin cells, fat cells and even beating heart muscle cells. This chaotic mess of cell types is not an organism. Cells kept in laboratory glassware must lack some necessary organizing factor that is present in the blastocyst. This factor may come from the blastocyst cells, which are not part of the inner cell mass, or perhaps the blastocyst structure holds the inner cell mass cells in the right spatial relationship to one another for orderly development to proceed.

In mice, which are very similar to people in their embryonic development, an embryonic stem cell can be tagged in various ways and injected into another mouse blastocyst. The mouse born from this blastocyst is a chimera, meaning an animal made of two different kinds of cells. Cells from the injected embryonic stem cell are typically found in every tissue of the mouse's body, establishing that all fates are open to it and that it is totipotent.

The equivalent experiment cannot, for ethical reasons, be performed in people. But Thomson injected human embryonic stem cells beneath the skin of a mouse and found that they formed cells belonging to the three main lineages of human embryos.

Could human embryonic stem cells be used to repair damaged human tissues? Researchers have already found ways to make mouse embryonic stem cells develop into heart muscle cells. When these were injected into another mouse's heart, they engrafted into the heart and behaved like the existing cells.[3]

In a startling demonstration of the power of embryonic stem cells to build new tissues, Ronald McKay and colleagues have made structures closely resembling the islets of the pancreas gland from the embryonic stem cells of a mouse.[4] In the pancreas, the islets secrete not only insulin but other hormones known as glucagon and somatostatin, each of which is made by a different type of cell. McKay and colleagues induced these and other types of pancreatic cell to form by a clever guess at the right factors needed to drive embryonic stem cells down the lineage to a pancreatic precursor cell.

To their surprise, they found the different cell types spontaneously assembled themselves into structures with the same cellular architecture as pancreatic islets. When exposed to glucose, the islet-like structures secreted insulin, just as real islets do.

The researchers then injected the islet-like structures under the skin of diabetic mice and saw the islets were able to induce a blood supply to form to them. The grafts did not cure the mice, probably because the cells were immature and not producing enough insulin, but the mice did live longer than usual.

If the experiment should prove to work as well in human cells as in mice, the basis would be laid for a possible novel treatment of type 1 diabetes, a disease that affects some sixteen million Americans and in which the pancreas does not produce enough insulin.

One reason for such confidence is that pancreatic islets transplanted from donor organs have been shown to work in a small number of patients, saving them the chore of daily insulin injections although they must take drugs to prevent the transplanted cells from being rejected. There are not nearly enough donor organs to meet the demand. Islets made from human embryonic stem cells could be produced in abundance, however, because the stem cells grow and divide indefinitely.

Another intractable disease for which embryonic stem cells hold promise is Parkinson's disease. Parkinson's is caused by the mysterious death of brain cells that produce dopamine, a chemical that conveys messages between neurons. For lack of the signals conveyed by dopamine, people cannot properly control their movements. The drug L-dopa works excellently at first, but its effects do not always last. For a time it seemed that Parkinson's patients could gain relief from injection into their brains of dopamine-producing cells extracted from aborted fetuses, but a carefully controlled study showed that the operation was not successful, at least in its present form.[5] Even if it had worked, there would never have been enough fetal cells to treat more than a fraction of the patients in need of treatment. Embryonic stem cells may prove more responsive to local cues and better behaved than the fetal cells.

One problem in treating patients with embryonic stem cells is the question of immune rejection, because the immune system generally attacks any cells perceived as foreign. Embryonic stem cells seem to be less antagonistic to the immune system than adult cells. Even so, it may be desirable to create a bank of embryonic stem cells so that patients can be treated with an immunologically compatible strain.

Another ingenious possibility is to rely on the embryonic cells' apparent ability to induce immune tolerance. In this scheme, yet to be tested, a physician seeking to give a patient a new pancreas, say, would take a sample of embryonic stem cells and convert some into pancreatic islet cells, others into the blood-forming stem cells of the marrow. First the blood stem cells would be injected into the patient and would take up residence in the bone marrow. There, they would induce the patient's immune system to tolerate them. Then the islet cells would be injected to reconstitute the patient's pancreatic function.

Creating Embryonic Stem Cells Through Cloning

Better still would be to take one of the patient's own mature body cells and walk it back through biological time until it was once again an embryonic stem cell. Can ordinary cells be converted to embryonic stem cells? Their genome is no different: all cells in the body have the same double set of chromosomes, except for the germ cells, which have one set, and red blood cells, which last for only six weeks or so and have lost their nucleus altogether.

The difference between an embryonic stem cell and a mature body cell presumably lies not in the genome itself but in the way the genome is controlled by the many proteins that attach to it. In a skin cell, say, whole suites of genes must somehow be permanently switched off, leaving active only general housekeeping genes and the genes whose products are essential for skin cells. The assignment of cell fate may occur as an embryonic cell receives certain signals from its neighbors and banks of genes are repressed, progressively narrowing the kind of cell it can become. A cell's fate is determined when all alternatives are denied to it and all cell-specific genes are switched off save its own.

The cloning of Dolly the sheep suggested that a mature cell's fate could be reversed and the cell reprogrammed to its embryonic state. Before Dolly, many animal species had been cloned from fetal cells, which retain more plasticity than mature cells. In the case of Dolly, the nucleus was removed from a sheep's udder cell and injected into another sheep's egg cell whose own nucleus had been removed. The fluid contents, or cytoplasm, of the egg cell presumably contained factors that entered the inserted nucleus and reprogrammed the genome of the mature udder cell back to its embryonic state. When the egg was transplanted into the womb of a foster sheep, it grew to be an almost exact replica of the sheep from whose udder cell it had come. Or so at least it is assumed—the udder cells used by Ian Wilmut were being grown in culture for other purposes, and their owner had long since become mutton.

Although the Dolly experiment has never been repeated with sheep, researchers have since cloned other animals, such as mice and cattle, from mature cells. Still, a curious feature of all these experiments is how difficult they are: regardless of the species, between two hundred and four hundred cell nuclei must be transplanted into egg cells for each successful cloning. Skeptics have suggested that what is being cloned in all these cases is not a mature, fully differentiated cell, but rather an adult stem cell, since stem cells might well exist in the adult tissues in approximately these proportions. Cloners, in other words, would be gaining apparent success only when they happened to pick a stem cell and not a differentiated cell.[6]

Assuming, however, that mature cells can be cloned, the way would be open in principle to create embryonic cells from a patient's own mature cells, thus enabling tissues to be regrown from his or her own cells and avoiding any problem of immune rejection.

There are three possible approaches, which together illustrate some of the strange roads and ethical frontiers that cell biology is testing. One approach, called therapeutic cloning, would be to take the nucleus from a patient's skin cell (or any other easily accessible cell) and reprogram it, as the animal cloners do, by inserting it into a human egg cell. This cell would not be implanted in a woman's uterus—that's the route to cloning a human, which is not at present ethically acceptable, although it may one day become just another specialist technique offered in fertility clinics. Instead, the egg would be allowed to grow and divide in laboratory culture; its inner cell mass would be removed and cultivated until there were enough embryonic cells to turn into whatever kind of tissue cell was needed for the patient's treatment.

The technical feasibility of therapeutic cloning has been demonstrated in mice by two Rockefeller University biologists, Teruhiko Wakayama and Peter Mombaerts. They took ordinary skin cells from mice (the usual way is to chop a tiny piece off the end of the mouse's tail) and converted them into embryos by transferring the skin cell nucleus into a mouse oocyte whose own nucleus had been removed. Embryonic stem cells derived from the embryos' inner cell mass were then converted into dopamine-producing brain cells of the sort that could be used to treat mice suffering from the mouse equivalent of Parkinson's disease.[7]

This kind of experiment has philosophically curious consequences because it means every cell of a person's body could in principle create a new individual. In therapeutic cloning, however, the embryos created this way would be kept in freezers as a permanent source of replacement parts throughout the owner's now doubtless lengthy lifetime.

But the human egg cells required by the technique are not so easily come by, and the procedure is too close for comfort to the vexing issue of human cloning. So a different, though hardly less startling, approach has been proposed by Advanced Cell Technology of Worcester, Massachusetts. The company's technical expertise was in the uncontroversial business of cloning cattle until it acquired a new chief executive in the form of Michael West, the visionary founder of the Geron Corporation of Menlo Park, California. It was at West's direction that Geron sponsored the research by Thomson and Gearhart that led to the cultivation of human embryonic stem cells, as well as the successful "immortalization" of human cells with telomerase.

West believed that there was an obvious way round the scarcity and ethical delicacy of using human egg cells to generate embryonic stem cells: cow eggs. Although that sounds like a recipe for generating a minotaur, the half-human, half-bull monster that inhabited the labyrinth in Crete, in fact a cow cell with a human nucleus would soon be taken over by the human genome and all the cow proteins replaced with human proteins, though with one important exception.

As it happened, hybrid cells of this type had already been made before West arrived at Advanced Cell Technology. One cell had divided five times and reached the stage where it seemed to have produced cells like those of the inner cell mass. At that point the experiment was stopped and apparently pursued no further. West decided to announce the experiment to test the ethical acceptability of exploring this line of research.[8]

The technical problems of using cow cells to create human embryonic stem cells are probably at least as severe as the ethical ones. Mitochondria, the energy-creating organelles that serve as the cell's batteries, differ from one species to another. As bacteria captured eons ago by animal cells, the mitochondria have their own small genome and produce some of their own proteins, but most of their proteins are produced by genes now resident in the cell's nuclear genome.

Because of this intimate interdependence between mitochondria and their parent cell, the mitochondria of one species are generally not compatible with the cells of any other species. In the Advanced Cell Technology experiment, the mitochondria in the cow egg's cytoplasm had evidently carried the hybrid cells through their first few divisions but would probably have proved incompatible with the human nucleus sooner or later. Injections of human mitochondria might have rescued the hybrid cell. And West, ever ingenious, spoke of genetically engineering herds of cows whose cells would be powered with human mitochondria; the eggs of these cows would serve as perfect incubators for human nuclei. So far nothing more has been heard of the cow egg–reprogramming technique, but it could resurface if other methods should fail.

A third approach is being pursued by Geron, West's former company, and that is to try to isolate the presumed factors in the egg's cytoplasm that are responsible for reprogramming the inserted adult nucleus. Geron has begun a collaboration with the Roslin Institute in Scotland, which owns the rights to the Dolly nuclear transfer technique. "The lesson of Dolly is that adult nuclei can be reprogrammed by cytoplasm," says Okarma, Geron's chief executive. "Our objective with Roslin is to learn the molecular biology of that process so that we can confer the reprogramming ability of cytoplasm to cells other than eggs."[9]

If Geron can identify the inferred reprogramming factors in the cytoplasm, it will gain a tool of extraordinary power in manipulating the human clay. By taking any ordinary cell of a person and exposing it to the factors, Geron could convert the cell's nucleus back to its embryonic state. The cells could be grown and perhaps stored, serving as a reservoir to replace each tissue and organ as the person aged. However, no progress on isolating the presumed factors has yet been reported.

Work on human embryonic stem cells holds high medical promise but at the same time is ethically fraught because of the way the cells are obtained and their closeness to the origin of an individual's creation. The National Institutes of Health, from which academic biomedical researchers receive most of their financial support, sought to fund research on the cells from the moment that Thomson first isolated them. Thomson himself used only private funds, from Geron, because Congress, through a temporary rule, had forbidden the NIH to support any research in which a human embryo was destroyed. Under its then director, Harold Varmus, the NIH in 1999 obtained a ruling from the Department of Health and Human Services saying, in effect, that it was OK for researchers to do experiments on human embryonic stem cells that others, such as Thomson, had isolated with non-government funds. But government-supported researchers could not themselves acquire and process blastocysts from fertility clinics.

The ruling did not pacify opponents of abortion, who viewed it as an effort to sidestep the specifically worded intent of Congress and, beyond the moral plane, were concerned about the political symbolism of the government seeming to condone the demise of human embryos. And ethicists were uncomfortable with the ruling's heavy pragmatism. To hold that researchers could use but not acquire embryonic stem cells sounded a lot like saying it was okay to use stolen property as long as you didn't do the stealing. Even the chairman of the National Bioethics Advisory Commission, a body generally favorable to medical research and to embryonic stem cell research in particular, said it was a "mistaken notion" to separate the derivation and use of the cells.[10]

The NIH nonetheless was able to finesse the congressional constraint on human embryo research and in August 2000 issued rules under which it would award research grants on embryonic stem cells. A year later President George Bush allowed government-supported research with human embryonic stem cells to proceed but only with cell lines created prior to August 9, 2001. Embryonic stem cells are perhaps less controversial than the vociferous debate in Washington might suggest. Despite the adamant position of the National Conference of Catholic Bishops, some Catholic theologians believe the moral status of the very early embryo prior to implantation is uncertain and that research may be permitted. "Within the Catholic tradition, a case can be made both against and for such research," says Margaret Farley, professor of Christian ethics at Yale University.

The Jewish tradition has no problem with using blastocysts for medical research. Early embryos "have no legal status whatever in Jewish law when they are outside the womb, because they have no potential for becoming a human being," says Elliot N. Dorff, a Conservative rabbi and author of a book on medical ethics. Medically beneficial research with the embryos is therefore "very much to be encouraged." [11]

But while the ethical and technical problems of embryonic stem cells are being worked out, another kind of stem cell research is coming to the fore.

Adult Stem Cells

Most of the body's cells are mature and fully differentiated. Their fate is cast. Some can grow and divide a little, but for the most part the power of change has been withdrawn from them. That makes sense from the point of view of controlling cancer. It takes only one of the body's 100 trillion cells to develop into a tumor, at any stage in the body's eighty-year lifetime, and the whole organism is put into peril. To help even the odds in the body's favor, cells are put under a wealth of constraints to deny them independent growth.

But several tissues of the body must constantly renew themselves. The entire skin is replaced every two weeks, as new cells pushing up from the lower layers replace the thousands of dead skin cells sloughed off each day. The cells of the stomach lining are rapidly worn out by the hellish conditions—high acidity, bacterial attack, and intense activity by protein-busting enzymes—that accompany digestion. They too must be constantly renewed. And the body loses, and must replace, about 1 billion red blood cells every day.

How do the skin, stomach, and blood renew themselves? Each tissue has a special class of source or stem cells that generate replacement cells while maintaining their own population numbers. These adult stem cells, also known as tissue specific stem cells, have a special trick known as asymmetric division. When ordinary cells divide, the parent cell becomes two identical daughter cells. In asymmetric division, a stem cell divides to produce one maturing cell and one stem cell. The maturing cell goes on to develop new tissue cells, while the stem cell remains available to create another tissue cell.

Because of asymmetric division, the stem cell population continually renews itself. Unlike embryonic stem cells, which have a fleeting existence, the adult stem cells are always around.

The stem cells of the blood, known as hematopoietic (blood-forming) stem cells, can generate a whole variety of different cell types, including those that make the oxygen-ferrying red blood cells and the white blood cells that constitute the body's immune defense system.

Blood, skin and stomach are not the only tissues to possess stem cells. It's now well established, despite years of firm belief to the contrary, that the brain too has stem cells that renew the neurons in at least two important regions of the brain. The stem cells provide a constant stream of new cells to the olfactory bulb, the part of the brain that holds the sense of smell, and to the hippocampus, an organ where initial memories of faces and places are formed before being transferred elsewhere.

There is even evidence that a stream of new cells arrives daily at the cerebral cortex, the highest level organ in the brain and the seat of consciousness. These daily deliveries of cells could be intimately connected with long term memory, since an obvious way to organize a temporal sequence of memories is to store each day's memories in a series of cells of sequential birth dates.[12]

As new types of adult stem cells came to light, it seemed that each tissue might have its own special reservoir of stem cells stashed away somewhere. Presumably this is why the human body retains such limited regenerative powers as it has. The liver can be partially regenerated. Wounds heal. Children under the age of ten can sometimes regrow the tip of a finger if the cut is above the last knuckle.

Recent findings suggest that adult stem cells from different tissues may have a lot in common with one another, since several types seem to be able to operate outside their own tissues. The hematopoietic stem cells, which reside in the bone marrow, also seem to be able to generate new muscle cells. New liver cells can be formed by cells of the bone marrow. The stem cells of the brain seem to be able to take up residence in the bone marrow and turn into hematopoietic stem cells.

These transformations are not yet understood, but they raise the question of whether all adult stem cells are pretty much interchangeable and just need the right signals from their surroundings to start repairing or replacing whatever tissue they are in. "Amazingly, when extracted from their niches and forced to take up residence at new body sites, it appears that some stem cells can adopt lineages previously not thought possible," say Elaine Fuchs and Julia A. Segre, two biologists at the University of Chicago.[13]

If adult stem cells are to some extent interconvertible, it will be much easier to find sources of such cells to treat a patient. Stem cells could be extracted from bone marrow and used to regenerate the brain or liver. Another accessible stem cell would be the skin's stem cells, now known to reside in a bulge halfway up the underskin part of the hair follicles that dot the skin.

Adult stem cells derived from a patient would of course raise no problem of immune incompatibility, as could occur with cells derived from a line of embryonic stem cells.

One instance of the adult stem cells' almost magical versatility is the ability of mesenchymal stem cells to take up different fates depending on their context. The cells are so called because they can build muscle, cartilage, bone, fat and tendon, all of which are tissues that derive from a sheet of cells in the early embryo called the mesenchyme. The cells can also make the stroma of the bone marrow, the cells that support the hematopoietic cells, and for that reason are also known as bone marrow stromal cells.

Most of these mesenchymal fates can be reproduced outside the body simply by providing the right physical cues for the cells. Researchers at Osiris Therapeutics, a company housed in a former tuna cannery on the Baltimore harbor waterfront, have found that mesenchymal stem cells grown on ceramic will turn into bone-making cells. If put in a gel, they will form cartilage. If injected into the bloodstream, they take up residence in the bone marrow and build stroma.

The company, named after the ancient Egyptian god of regeneration and immortality, already has one product in clinical trials, an infusion of mesenchymal cells for patients undergoing bone marrow transplants. The company hopes the cells, by building stroma for the hematopoietic cells that are injected later, will enable the blood system to be reconstituted faster. Osiris also has a bone-making preparation of stem cells, OsteoCel, in clinical trials, and is working on animal experiments to develop cartilage-making cells for joint repair and heart muscle cells for heart attack victims.

In principle, mesenchymal cells should be able to repair all the body's connective tissues—the bone, muscle, tendon and cartilage that are most subject to damage as the body ages or is wounded. If the cells work as well as hoped, says Osiris's chief executive, Annemarie Moseley, "Patients in middle age would come in, before the degenerative process starts, and would get a cell or tissue implant, and degeneration of the structural elements would no longer be part of the aging process."

Two dramatic advances have been reported in the use of adult stem cells to treat heart disease, a development of considerable importance because the human body does not make new heart muscle cells. People must make do with their allotment at birth. In a heart attack, when an artery supplying the pumping heart muscle is blocked off, the heart muscle cells that die for lack of oxygen are not replaced. The heart forms a scar around the dead tissue and, if the patient is saved with a bypass of the blocked artery, will continue beating though with less efficiency than before. But if new heart muscle cells could be injected into the heart, the organ could perhaps be restored to much nearer its original state.

In one of the new studies, reported in April 2001, researchers at the New York Medical College in Valhalla, New York, and the National Institutes of Health purified stem cells from the bone marrow of mice. They were careful to select the most primitive hematopoietic stem cells, as the blood-forming cells of the marrow are known, ones that had not yet taken even the first steps toward becoming a red or white blood cell.

These hematopoietic stem cells were then injected directly into the heart muscle of mice that had artificially been given a heart attack by tying off one of the heart's arteries. As if taking the proper chemical cues from their new surroundings, the stem cells matured into the three principal cell types found in heart tissue—heart muscle cells, the smooth muscle cells of the artery walls, and the cells that line the inner surface of the blood vessels.[14] "The implications are profound: damage to the heart muscle after a heart attack might be reparable with specific bone marrow cells," wrote Mark Sussman, a cardiac researcher, in a commentary on the article.[15]

The finding is so much on the frontier of research that its significance is at present hard to interpret. It is an almost complete surprise that hematopoietic cells should be able to morph into heart tissue, although there have been other hints of the cells' surprising versatility. Could it be that the bone marrow is the long sought site of the heart's stem cells? Is the long-standing dogma that no new heart muscles are formed after birth incorrect? And if new heart muscle cells are in fact regularly derived from their secret source of bone marrow stem cells, why does this replacement system not spring into action after heart attacks?

A second report, from researchers at Columbia University, showed how the heart muscle of rats given heart attacks could be repaired with a novel kind of stem cell the researchers had discovered in the human bone marrow. The role of the cells, named angioblasts, is to generate the cells that make the body's fine blood vessels.

Stem cells are much less provocative to the immune system than mature cells, a fact which enabled the researchers to inject human angioblasts directly into the rats' bloodstream.

Remarkably, the cells homed in on the damaged tissues of the rats' heart because the tissues were sending out a distress chemical to advertise their dangerous lack of oxygen. The angioblasts formed a new blood supply which delivered oxygen, saved the heart muscle cells from dying, and prevented the scar tissue that usually forms after a heart attack.[16]

Could heart attacks be treated by the simple expedient of injecting the patient with angioblasts? To pursue the idea, the Columbia physicians intend to start clinical trials after further experiments in rats to settle questions of efficacy and dose.

The Stem Cell Revolution

Clearly both kinds of stem cell, embryonic stem cells and adult stem cells, hold significant promise for treating a wide range of diseases, many of them otherwise incurable. Opponents of abortion have argued that there is no need to use embryonic stem cells when adult stem cells would suffice. But many biologists believe it is necessary to explore the advantages and disadvantages of both kinds of cells in order to make the best medical choices.

Stem cells of both types promise to open out a new dimension of medicine in addition to their own inherent properties, and that is the promise of genetic engineering that gene therapy has so far failed to deliver.

Once stem cells are outside the body and before they are inserted into the patient, they can be manipulated in various ways. If a patient suffers from a genetic disease that affects a particular tissue, the stem cells for that tissue could be treated with the correct form of the gene. Genes often do not insert themselves correctly, but because large numbers of stem cells can be grown outside the body, the rare cell where the new gene has integrated itself correctly into the DNA could be identified and cloned.

Most forms of gene therapy depend on viruses to deliver corrective genes to the body's cells. Viruses are convenient vehicles because in the course of evolution they have mastered the art of entering human cells, penetrating the nucleus and integrating their genetic material into the

DNA of the chromosomes. In gene therapy, a virus's disease-causing genes are removed and replaced with a medically beneficial gene. But though the idea is fine in principle, in practice it has seldom worked as advertised. The virus does not insert the new gene in the same place as the defective one, so it is not governed by the proper control mechanisms, which are special sequences of DNA that lie in front of a gene and control how and when the gene is expressed.

Another persistent problem with gene therapy is that the body's immune system is expert at spotting and destroying all cells that harbor a virus within. Thus even if a beneficial gene has been inserted by a virus, the cell is doomed because of the virus's presence; after six weeks or so all the altered cells have been destroyed.

Stem cells may offer a better way of delivering corrective genes into the body because biologists can insert genes into a population of stem cells outside the body and select the one cell in a million in which the genes have inserted correctly. That cell can then be grown into a clone of genetically repaired cells for implantation in the patient.

Another thing that biologists may do with stem cells before implantation is to "immortalize" them. If mature cells are cultured in the laboratory, they can be made to divide about fifty times but will then lapse into senescence, as if some finite thread of life had come to an end. This division constraint is known as the Hayflick limit after Leonard Hayflick of Stanford University who discovered it in 1961.

The physical basis of the Hayflick limit appears to lie at the tips of the cell's chromosomes, where the same six-nucleotide sequence is repeated thousands of times over. The tips of the chromosomes, called telomeres (meaning end sections), get shorter each time the cell divides; when the telomeres reach a certain minimum length, the cell can divide no more and it enters senescence.

The telomeres must of course be lengthened in some way or infants would have shorter telomeres than their parents and the species would quickly fall extinct. The genome includes a gene that makes telomerase, an enzyme whose role is to extend the telomeres by building back extra copies of the six-nucleotide sequence. Telomerase is fully active in the egg and sperm cells, ensuring that all the body cells of new-borns have long telomeres. But in mature cells the telomerase gene is permanently switched off.

The human telomerase gene was first identified in 1997 and the license to it was snapped up by Michael West, Geron's founder and then research director. Research groups at Geron and the University of Texas inserted an copy of the telomerase gene into human cells. Cells with this extra copy, which unlike their own suppressed gene was in active form, were able to grow and divide an indefinite number of times beyond the Hayflick limit, the researchers reported in January 1998.

The breaking of the Hayflick limit was a dramatic result that confirmed the idea of a built-in constraint on the number of divisions permitted to a mature cell. But what was its practical significance?

The finding would "now allow us to take a person's own cells, manipulate and rejuvenate them and give them back to the same patient," says Jerry W. Shay of the Texas group, suggesting that the telomerase technique would be of particular help for burn victims and people with diseases caused by the failure of aging cells to divide, such as macular degeneration. On the strength of this news, Geron's stock price rose 44 percent.

But the skeptics were quick to criticize Geron's grounds for optimism. Evolution had surely not gone to the bother of engineering the elaborate telomere-telomerase system just to limit human life. It must serve some more useful purpose, and the obvious application is as a last ditch defense against cancer.

If any cell should escape the usual constraints on free division and start to multiply on its own, it could divide fifty times or so, enough to make a microscopic tumor, and would then hit the wall of the Hayflick limit as its telomeres reached the critical minimum length. Most cancers do indeed have an active telomerase gene, suggesting that this is yet another of the cell's control mechanisms that must be subverted for a cancer to take hold. If telomeres are an anti-cancer mechanism, not a life-shortening mechanism, it would be positively dangerous to equip cells with an active telomerase gene. "Geron would have us believe that telomerase is the key to immortal life, and I have no idea if there is any wisp of truth in that," says Robert Weinberg, a leading expert on cancer genetics at the Whitehead Institute in Cambridge, Massachusetts. No one, to his knowledge, has ever died of short telomeres, so he feels that longer telomeres are unlikely to promote longevity.

Geron's scientists agree that the primary purpose of the telomere system is cancer protection but believe that human cells may indeed run out of telomeres in certain circumstances, such as when tissues are put under unusual stress requiring incessant multiplication of cells. In people with cirrhosis, for example, the liver cells have been found to have unusually short telomeres. It may be that their liver cells are called on to divide so often, because of the daily destruction caused by alcohol, that they run out of telomeres, at which point the liver can no longer regenerate and scar tissue forms instead. Shortened telomeres may also play a role in several other diseases, including AIDS, where the T cells of the immune system are depleted by continual assault by the AIDS virus.

Geron scientists believe that such tissues could be rejuvenated by having their cells' telomeres rebuilt to their youthful lengths. The treatment would consist of a onetime application of telomerase to restore the telomeres, not a permanently active telomerase gene that would sabotage the cells' anti-cancer protection.

Many stem cell biologists believe that they are on the verge of being able to use the cells in medically significant ways. From looking at the DNA in the genome, as most molecular biologists do, it is hard to predict the behavior of any individual cell. Hence a key feature of cells has been ignored, says McKay of the NIH, and that is that the body is designed as a self-assembling system. Given the right cues, cells will organize themselves into the tissue or structure appropriate for that position in the body.

"It's incredibly difficult to translate the information in the genome to the level of the whole organism simply by doing genetics," McKay says. "But above the level of the central dogma, cells are self-assembling. The consequences for cell biology are massive. That's been missed by generations of scientists who have been trained as chemists. We are so heavily trained to think in a reductionist way. That is why we are blinded to the beauty of the system."

The beauty of the system, the forest invisible to the molecular biologists focused on the trees, is in McKay's view the conversation among the body's cells that organizes their three-dimensional structure and dictates the size and placement of each of the body's organs. "The liver stays a constant size," he says. "There has to be space. Animals are composed in terms of space—a dynamic equilibrium of cells signaling to each other. It's a society of cell types, all communicating. It's a harmony of cells. That's why when we put cells into animals they become part of the normal order of the system.

"I anticipate that in two years it will be obvious that this is the field where huge resources should go, and in five to ten years it will have a major clinical impact. Diseases that are significant causes of death will be impacted, and that will impact life span," McKay declares.[17]

Elaine Fuchs and Julia Segre, two stem cell biologists at the University of Chicago, say the potential uses of stem cells seem endless, provided that researchers can learn how to make the cells differentiate down the desired pathways. These uses include, in their view, "the generation of different types of neurons for treatment of Alzheimer's disease, spinal cord injuries, or Parkinson's disease, the production of heart muscle cells for congenital heart disorders or for heart attack victims, the generation of insulin-secreting pancreatic islet cells for the treatment of certain types of diabetes, or even the generation of dermal papilla or hair follicle stem cells for treatment of certain types of baldness."[18] It is not often that serious researchers promise a cure for Alzheimer's, Parkinson's, heart disease, diabetes, and baldness in the same sentence.

Much the same list is offered by Mary Hynes of Renovis and Arnon Rosenthal of Genentech. "The prospect of greater longevity, associated with an increased expectancy of good health, underlies the intense excitement surrounding work on embryonic stem cells," they say, since the cells may provide "replacement body parts in cardiovascular, autoimmune, Alzheimer's, or Parkinson's disease or for diabetes, osteoporosis, cancer, spinal cord injury or birth defects."[19]

Regenerative Medicine

One element that is missing in the general euphoria over stem cells is knowledge of the signals from the surrounding milieu that cue the cells to differentiate down specific paths. Some of these signals are already known, but until many more are identified and decoded, biologists will not fully understand the language by which the body's self-assembly is governed.

This is where the human genome sequence is likely to prove useful. Both the signaling proteins and the receptors that detect them must be exported from cells, and exported proteins have a special sequence of amino acid units that is recognized like a zip code by the cell's protein-sorting mechanism. (The receptor proteins are not fully exported, just embedded in the cell's outer membrane, but they nevertheless have the same general zip code as the fully exported signal proteins.)

The DNA sequences that encode the protein zip codes can be identified in the genome. Celera, in its initial annotation of the human genome, found some 1,500 signaling and receptor genes. William Haseltine, chief executive of Human Genome Sciences, says that by the EST method—direct capture of gene transcripts made by cells—he has found some 11,000 signaling and receptor genes.

Whichever number is right, the way seems open to exploring the communications system of the body and in particular the subset of signals that drive stem cells to their different fates.

Haseltine sees signals and stem cells as two principal components of the regenerative medicine of the near future, since if the correct signals are applied to stem cells in the right sequence, the cells can be made to develop into any desired tissue of the body.

He estimates that there are about two thousand different types of cells in the body and that each is derived from the original embryonic stem cell by the action of five to ten different signaling proteins. These signals are produced by neighboring cells as part of the embryo's self-assembly routine. So by tracking which genes are switched on in each cell type at each stage of fetal development, it may be possible to de-construct the program of signals that causes all the different organs and tissues to develop in the appropriate relationship to one another.

Haseltine's company, Human Genome Sciences, has captured gene transcripts from fetal tissues at successive stages through the first twelve weeks of life. "We find 80% of the genes stay the same and 20% change. That's one reason why we got a more complete picture of human genes—many genes are expressed early in life and not later," he says.

"For the first time most of the key signaling molecules are at hand," Haseltine said in a talk about regenerative medicine. "A new frontier is dawning. . . . If successful, we should be able to re-build and make new virtually any tissue or organ in the human body that is damaged or diseased. That is what I call the promise of regenerative medicine." [20]

The concept of regenerative medicine raises enigmatic questions about human longevity. Suppose it should become possible to replace or rejuvenate each organ as it becomes old and inefficient. With the ability to grow a new heart, new lungs, new liver, preferably from their own stem cells, what will people die of? Even the brain, long assumed to be composed of non-replaceable cells, has been found to possess stem cells that routinely renew the cells in at least two of its component parts. So some parts of the brain can be included in the list of renewable body parts. What then will give out? Is there a necessary weak link or could the human body be kept going indefinitely?

Regenerative medicine, if it works, could have two kinds of effects. First, it could increase life expectancy, enabling more people to attain their full natural life span and to enjoy good health while doing so. Second, it could increase life span itself.

The first possibility, increasing life expectancy, lies fully within the agreed goal of contemporary medicine. Increasing the life span is a different and far more ambitious goal. Okarma, chief executive of Geron, says firmly that his concept of regenerative medicine does not include increasing the natural human life span. "Our objective is to increase the health span, not the life span," he says. "Our hope would be that our children will live a greater fraction of their life span in wellness. I suppose that in theory you could take an individual and continually rebuild that person with the technologies we have been talking about. But that would be an extreme case. I would not be easy if this ends up making superpeople who can afford to have this technology applied to them multiple times."[21]

But Michael West, Geron's founder, believes the goal of the new medicine need not stop at life expectancy. "I expect this to extend life span," he says, "though I can't say by how much. Cars have a finite life span, but you can replace each component with new components; theoretically you can extend the life span of a car indefinitely. To really affect the human life span, we would have to affect the vascular system and the central nervous system, but I think it is possible."[22]

Can human life span be increased? That depends on the nature of aging and whether evolution has shaped the life span in a way that is easy for biologists to manipulate.

6
The Quest for Immortality

A most unusual conference took place recently in the museum of the University of Pennsylvania. Its subject was immortality. Prompted by advances in extending the life span of laboratory animals, the conference was convened to consider the consequences of people too being able one day to live a very long time.

The scientists at the meeting said, in effect, that life span was clearly a phenomenon under genetic control, so they expected to be able to increase it just as soon as they laid their hands on the right genes.

Theologians and ethicists did not greet the news with open rapture. The Reverend Richard J. Neuhaus, of the Institute on Religion and Public Life, termed the search for immortality "a pagan and sub-Christian quest" driven by the "essentially amoral and mindless dynamic of the technological imperative joined to an ignoble fear of death." [1]

The ethicists found the idea of eternal life no more appealing. "To argue that human life is better without death is to argue that human life would be better without being human," declared Leon R. Kass, a biomedical ethicist at the University of Chicago. He praised the stalwart character of the Homeric hero Odysseus, who turned down the offer of immortality from the nymph Calypso, preferring to live out a normal life with his aged wife, Penelope. "The finitude of human life," Kass declared, "is a blessing for every individual whether he knows it or not."

His stern views were echoed by Daniel Callahan, an ethicist at the Hastings Center in Garrison, New York, who opposed not just immortality but even medicine designed to increase life expectancy. Proposals to increase life expectancy should be assessed "in the context of enormous skepticism since no one has shown the present life expectation is somehow a disaster," he said. "We can't ban this research but we can make it socially despicable."

As a philosophical joke, the organizers had convened the conference in the museum's Egyptian room. Speakers read their notes standing almost beneath the nose of a large sphinx and surrounded by accoutrements of the Egyptian cult of immortality. Five thousand years after the pyramids, scientists had at last arrived at a real prospect of being able to extend human life, only to hear themselves denounced for their pains as pagans engaged in a sub-human quest. If the mummies in their shrouds could have listened, they would surely have wept. Almost the only words in favor of prolonging life were uttered by a rabbi, Neil Gillman of the Jewish Theological Seminary. "I can violate the Sabbath and Yom Kippur even if they are on the same day in order to preserve life," he said.

Only a few years ago, no one would have bothered to turn up for anything so ludicrous as a discussion of immortality. The topic of rejuvenation had been lastingly discredited by clinics that offered various animal gland extracts in the 1930s. Research on aging was a backwater from which little progress was expected. Human mortality statistics showed no sign of deviating from the iron-clad rule worked out by Benjamin Gompertz, a British actuary, in 1825, that after puberty the human death rate doubles inexorably every eight years; the older you get, in other words, the more likely you are to die.

Evolutionary biologists had figured out a gloomy explanation for death, that natural selection favors any gene that helps an animal survive up to the age of reproduction but not beyond; moreover, so many genes are involved in senescence that there would be no hope of counteracting all of them. And the Hayflick limit, the 1961 discovery that human cells kept in glassware could divide only fifty times or so before dying, seemed to establish that the basic clay of the human body had a finite lifetime.

Today, these dismal limitations appear distinctly less iron-clad in the light of three remarkable findings. First, people in developed countries are living much longer than before and from these older populations has emerged a remarkable defiance of Gompertz's gloomy law. It seems that once you make it to a certain age, around seventy-five to eighty, your chances of dying start to decelerate.[2]

Life expectancy has steadily increased, from perhaps twenty-five years for much of human history, to fifty years by 1900 in the most developed countries, and has now reached eighty years in Japan and elsewhere. The number of Americans over one hundred years old almost doubled, from 37,000 to 70,000, during the 1990s, and is projected to hit 1 million by 2050.[3] In England, where it is the custom for the sovereign to dispatch a birthday telegram to subjects attaining their hundredth birthday, the queen dispatched only 255 telegrams in 1952 but 5,218 in 1996.[4]

This dramatic increase in life expectancy, which one demographer calls "arguably the greatest collective human achievement,"[5] has presumably been brought about by public health measures, antibiotics, various medical interventions, and better nutrition. A rather surprising feature of the increase is that it has shown no sign so far of leveling off. Maximum life spans have also been rising steadily. For example, the maximum reported age at death for Swedish men and women has risen steadily for the past 130 years that good records have been kept. This means it is hard at present to say what the limit of life expectancy may be or what the maximum human life span is.

The longest well-attested life span was that of Jeanne Calment, who died on August 4, 1997, at the age of 122 years and 5 months. Calment attributed her longevity to a diet rich in olive oil, regular glasses of port wine, plenty of exercise, and a sense of humor. Geneticists favor the possibility that she carried some interesting genes. Of her fifty-five immediate ancestors, 24 percent lived to eighty years or more, compared to only 2 percent of a comparison group of people. "One might hypothesize that the extraordinary longevity of Jeanne Calment is largely genetic in origin and that it is due to an exceptional genetic inheritance, randomly accumulated within the ecological niche of Arles in the 18th and 19th centuries (and, more specifically, within the social group of craftsmen and shopkeepers running prosperous businesses in the town)," say two French demographers who studied the ages and occupations of her forebears.[6]

The second blow to the doom and gloom view of human mortality concerned the Hayflick limit. "As the most frequently cited result in all of gerontology, the Hayflick limit has contributed powerfully to a general sense that the study of longevity is a study of limits, trade-offs, and diminishing returns," a demographer wrote in 1997.[7] One year later, the Hayflick limit was broken. Human cells given an active copy of the telomerase gene, as recounted in the last chapter, proved able to grow and divide an indefinite number of times beyond the fifty or so divisions that would otherwise have been their limit. Though the natural role of the telomerase system is still under study, the breaking of the Hayflick limit undermined the concept that death is inevitable because of some inherent weakness in the living cell. Cells can grow and divide indefinitely—they are "immortalized," in biologists' parlance—if the natural constraint of the telomerase system is reversed.

A third blow to fatalism has been a growing number of experiments in which laboratory animals—the C. elegans roundworm, the Drosophila fruit fly, and the mouse—have been made to live far beyond their usual lifetimes by the manipulation of a single gene. To evolutionary biologists, this is an extremely surprising development because their theory predicts that life span is determined by a very large number of genes and so should not be significantly altered by changing just a single gene. Almost every gene is likely, in one way or another, to affect an animal's fertility and its survival to the age of reproduction, and to be fine-tuned in shaping both so that the animal can leave the greatest number of progeny. "Everything that we know about the evolution of aging suggests that it is probably the most polygenic of all traits," writes Linda Partridge, an evolutionary biologist at University College, London. "Fertility and viability are affected by all genes in the genome. The prospects for genetic intervention, therefore, must be explored with some circumvention."[8]

These three developments—the defiance of the Gompertz law, the breaking of the Hayflick limit, and the genetic extension of life span in animals—have helped bring about a decisive shift in mood, from a belief that death is inevitable and immutable to a state of optimism, maybe excessive, in which it seems at least worthwhile to hold a conference on immortality.

Within the field of biological research, there has developed an interesting and maybe fruitful tension between evolutionary biologists, whose work on aging is largely though not entirely theoretical, and molecular biologists, who prefer to decide matters by experiment, not theory.

"Aging is a very tricky phenomenon, particularly if you don't understand evolutionary biology, which no cell biologist does," says Michael R. Rose, an expert on the evolution of aging who works at the University of California, Irvine.[9] Gary Ruvkun, a geneticist who works with roundworms at the Harvard Medical School, harbors a similar degree of respect for evolutionary biologists. "I've never bought into these theoretical arguments and regard them as *Just So Stories*. I can make up my own *Just So Stories*," he says, referring to Rudyard Kipling's children's tales and their ad hoc explanations of how the leopard got its spots and the rhinoceros its skin.[10] "The reason I don't like to engage too much in that kind of discussion is that I like to do experiments, and theoretical biology tends to be a little bit boring," says Leonard Guarente, a geneticist at the Massachusetts Institute of Technology.[11]

The differences between these two schools of thought are worth exploring because they frame the possibilities for increasing human life span. The evolutionary biologists' explanation of aging rests in large part on ideas developed by the late Peter Medawar, an immunologist at the National Institute for Medical Research in London, and George Williams of the State University of New York at Stony Brook.

Medawar's insight was that the power of natural selection drops sharply after the reproductive years. A person born with a genetic disease such as Werner's syndrome, which causes advanced symptoms of aging and early death, will not leave any descendants. Natural selection speedily eliminates all early-acting fatal genes from the population. When a new case of Werner's syndrome occurs, it is because of another spontaneous mutation in the same gene. But against late-acting deleterious genes, such as those that cause cancer, heart disease or Huntington's chorea, natural selection can do little because individuals with these variant genes have already had their children by the time that the gene's adverse effects are felt.

And just as natural selection is powerless to eliminate late-acting bad genes, it also has little power to foster the increase of late-acting good genes, such perhaps as those that might prevent cancer. The reason is that people with good late-acting genes are likely to be rather few in number, and their progeny will be far outnumbered by those of people who lack the gene. So natural selection has no leverage with which to make genes kindly to nonagenarians become more common in the population.

The intensity of natural selection depends a lot on the degree of hazard in the animal's surroundings. If you are a small, vulnerable species living in a forest with many relentless predators, natural selection will show a decided preference for genes that help you mature quickly and breed prolifically. But if you are large and fierce and live amid plenty, natural selection no longer places such a premium on rabbitlike fecundity. It gains the power to select for later-acting beneficial genes, such as those that give your species a stout heart and strong immune system.

Most mutations are harmful—not surprisingly, because each protein has been selected over millions of years to perform a precise function, so that a mutation, like any random change in a delicate piece of machinery, is more likely to disrupt its function than enhance it. Imagine, then, a steady trickle of mostly bad mutations occurring generation after generation in the human genome. The early acting ones will be screened out because their owners tend to have few or no progeny; the late acting ones will accumulate, bearing an inexorable burden of disease and degeneration. A variation of this dismal theme, developed by Williams, notes that many genes have more than one effect in the body (through one of nature's economies, the same gene may be used at different stages of life to do different things). Suppose a multi-role gene has an early acting effect that mildly benefits fertility and a late acting effect that promotes serious heart disease. Natural selection will favor such a gene because the slight benefit to fertility outweighs the devastating cost to longevity on the only scale that evolution cares about—the number of progeny.

The evolutionary theory of aging, if true, sheds an interesting light on the nature of death. Far from being a necessary end, death is purely contingent, a tail-end consequence of other more important processes. People decay and fall apart not because the flesh is weak but because natural selection has not had the opportunity to select for genes that protect the body in age.

Two hallmarks of a good theory are that it should explain something interesting and should make testable predictions. The evolutionary theory of aging does both. It helps explain the very different life spans enjoyed by different species. A small and vulnerable animal such as a mouse faces many lethal hazards—owls, foxes, and freezing nights. Mice in the wild live about four months. Given such a hazardous environment, it would be a waste of resources to design a mouse that lived a hundred months; it would never get to see 96 percent of them. Better to channel effort into early breeding, not longevity. And indeed evolution has designed mice to live in the fast lane, reaching breeding age early and having as many progeny as possible before the barn owl strikes.

An elephant, on the other hand, is too large to have any natural predators, except people. It does not freeze to death on cold nights. Older animals look after their infants, so there is an adaptive advantage (meaning one favored by natural selection) in having mothers survive many years after the birth of their offspring. And elephants are social animals led by a matriarch on whose experience the group of related animals depends. All these factors combine to make longevity an asset, and indeed elephants have evolved a maximum life span of more than seventy years.

Being genetically controlled, the life span of a species is not forever fixed but adapts according to circumstance. Two striking proofs of this idea exist, one furnished by experiment, one by nature.

Michael Rose has performed a well known series of experiments on the *Drosophila* fruit fly in which he created an advantage for longer lived flies either by choosing only the longest lived females in each generation to be the mothers of the next, or by destroying all eggs produced by flies less than a certain age. By these means he has created strains of flies whose longevity is almost double that of the original population. How does this happen? Presumably the owners of those late-acting beneficial genes now have proportionately more progeny and their genes have a greater chance of being represented in the next generation. In a surprisingly few generations longevity has increased. In one experiment the average and maximum life spans of the females rose by 30 percent after just fifteen generations of selective breeding, while the males' life spans rose by 15 percent.[12]

A beautiful natural experiment that made the same point was performed, or rather discovered, by Steven N. Austad, an expert on aging at the University of Idaho. He noticed that the opossums he was trapping at a field station in Venezuela showed severe signs of senescence at a mere two years of age. Opossums are small, defenseless creatures with many predators. According to the evolutionary theory of aging, the opossums faced accidental death so often that natural selection could not favor individuals whose genes enabled them to live longer. So if the mortality rate were reduced, shouldn't the opossums live longer?

Hoping that nature might somewhere have performed the experiment to answer this question, Austad looked for an island free of predators, with a warm climate and its own long-established opossum population, and not so close to the mainland that opossums could swim across and exchange genes. After much searching, he located the opossum Garden of Eden on Sapelo Island, a dagger-shaped patch of land five miles off the coast of Georgia and inhabited by small deer, large rattlesnakes, many ticks, and opossums.

The Sapelo opossums knew they were in paradise. They slept on the ground, not in burrows. They walked abroad at midday. They had lost the fearful adaptations of their mainland ancestors. And just as theory predicted, they had already begun to age more slowly. They lived on average 25 percent longer than mainland opossums, with a 50 percent greater maximum age. They had fewer pups per litter but a much greater likelihood of breeding a second year. Their tissues, as judged by the collagen in their tail tendons, aged more slowly. Austad concluded, "Thus by every measure I had examined—mortality rates, reproductive decline, and tendon aging—these island opossums were aging more slowly, just as the theory suggested they should."[13]

Implicit in the evolutionary theory of aging is the idea of a trade-off between fertility and longevity. Given finite resources, an animal can invest either in breeding prolifically or in maintaining its tissues and immune system so as to live a long life. This trade-off is genetically designed into the animal's life history. The Sapelo Island opossums lived longer but had smaller litters. In men the trade-off is mediated on the physiological level by the sex hormone testosterone, as is known from a study of longevity records of inmates in an unnamed Kansas institution. In the 1960s it was common practice to castrate mentally retarded people in institutions. Compared with other inmates, the castrated men lived almost fourteen years longer on average, and the gain in longevity was greater the younger the age at which the operation had occurred.[14]

The evolutionary theory of aging helps account for the vast range of life spans seen in nature. Each species' life history is adapted to its circumstances, with degree of hazard being one major influence. The basic clay of which all living creatures are made—the cell—is not a constraining factor. Cells will support lifetimes of days or of centuries, whichever the organism requires.

From birth to death, the *C. elegans* roundworm enjoys just twelve to eighteen days of life. But some animals live so long that their maximum life spans are unknown. It is not clear that sea anemones ever die; they have been kept in tanks for ninety years with no evidence of senescence. The ages of seashells can be assessed from their annual growth rings. Here the winner is the ocean quahog, some of which have been counted with 220 annual rings. Even the freshwater mussel can live 120 years.

The longest lived organisms, at least those whose ages can be measured, are trees. Redwoods and certain other conifers can live from 300 to 1,500 years. The oldest known trees are bristlecone pines, which can live at least 5,000 years, as measured by counting their annual growth rings.[15]

Does the theory of aging preclude the possibility of lengthening the human life span? The answer seems to be no. Evolutionary biologists do not expect it to happen, because so many genes are involved. Even if by some means life span were to be extended, some serious trade-off could be expected in the form of a sharp reduction in fertility. But their bottom line is that life span is not fixed. The thread of life is extensible, even if only in principle.

"There aren't any fixed limits to the lifespan of a particular species," says Rose. "Aging is evolutionarily tunable, and thus it must be genetically or biochemically malleable as well."[16]

S. Jay Olshansky, a demographer at the University of Chicago, has reached a similar conclusion. "If senescence is in fact the product of evolutionary neglect rather than evolutionary intent, then there is every reason to be optimistic that the process is inherently modifiable, an extremely important implication for an aging population," he and his colleagues say.[17]

So much for theory; what can researchers do in practice about longevity? There are two kinds of intervention that can extend life span, though so far only in animals. One is diet and the other is genetic manipulation.

The Wonders of Caloric Restriction

The diet experiments, first begun in the 1930s, consist of feeding rats or mice diets that have 30 percent fewer calories than when the animals can eat as much as they want. These are not malnutrition diets; they have all necessary vitamins and nutrients, but far fewer calories. The effects on life span are dramatic: the average life span of white rats is twenty-three months, but on a lifelong calorically restricted diet it soars to thirty-three months, an increase of 43 percent.

Richard Weindruch of the University of Wisconsin, who has performed many of these feeding experiments, finds that even if the calorically restricted diet is not started until early middle age, the maximum life span of mice is still extended by 10 to 20 percent.[18]

Animals kept on such a spartan diet generally weigh two thirds or even half the weight of their normally fed counterparts. But they are wonderfully healthy. They have better immune systems and somehow postpone most of the diseases common in later life, particularly cancer.

There is one downside. Just as evolutionary biologists would predict, fertility declines. With both rats and mice, puberty is delayed by caloric restriction. The adult rats are less prolific. As for the mice, they cease to breed at all. Their Darwinian fitness has been reduced to zero.

Some biologists have expressed skepticism about these feeding experiments. Laboratory mice are considerably overweight compared with their wild brethren; perhaps the calorically restricted diet is making them healthier simply by forcing them to revert to a diet more similar to what they enjoyed in the wild. These discussions seem to have been largely resolved in favor of the idea that caloric restriction does indeed increase life span in rodents, raising the questions of why this should be so and whether it would work in people.

Many animals seem to have an ancient strategy programmed into their genomes by the vicissitudes of existence. The strategy is simple: when times are hard and food is scarce, postpone reproduction and channel the body's resources into outliving the present famine. When food is scarce, the *C. elegans* roundworm, which usually lives two to three weeks, slips into a special hard-skinned larval state in which it can survive for many months. The response of mice and rats to a calorically restricted diet may represent a similar survival strategy.

Do people have the same response programmed into their genes? To address the question in primates, George S. Roth of the National Institute on Aging has been conducting a calorie restriction study on rhesus monkeys since 1987. Another group of monkeys is being studied at the University of Wisconsin. Because monkeys' natural lives in captivity are so long, definitive results are not yet available. But the evidence so far is promising. As of September 2000, there had been fewer deaths among the dieting monkeys, who receive 30 percent fewer calories than a comparison group of normally fed monkeys. The mortality difference is not yet statistically significant, Roth says, but the restricted monkeys have already developed metabolic patterns that suggest they will be more resistant to diabetes and heart disease.[19]

But even if the monkeys do show the same gains in longevity as do rats and mice, which would suggest that humans could extend their life span too, probably very few people would be able to benefit. A restricted calorie diet is extremely hard to follow. More than half of Americans are overweight, and of these 19 percent are obese, despite the cost in health and longevity; probably not so many could summon the willpower to maintain a diet with 30 percent fewer calories than normal, stifling their hunger pangs at every step past the bounty of the supermarket aisle.

But what if there were a way to achieve the benefit of a 40 percent longer life without the pain of semi-starvation? What if you could simply take a pill to trigger the longevity effect and not have to count calories at all? Such a pill, should it do to people what caloric restriction does to rodents, could have a major social impact: a 40 percent increase in longevity would have the average person in developed countries living to age 112 and the maximum life span raised to 168 years.

It's not so wild a prospect. As molecular biologists probe the triggers of aging in laboratory animals, they are perhaps drawing closer to the genes that trigger the fertility/longevity trade-off.

Weindruch and a colleague at the University of Wisconsin, Tomas A. Prolla, have gained a deep insight into the nature of aging at the level of the cell. In what may have been one of the first applications of genomic data to the field of aging research, they used an Affymetrix gene expression chip programmed with 6,000 mouse genes of known function. Mice probably have around 30,000 genes, so the chip included a large sample of the mouse's working parts.

A gene expression chip is designed to signal when any of its component genes are switched on in the cells under test; the chip can also quantify how actively the gene is being used. Weindruch and Prolla first looked at muscle cells from normally aging mice and then compared them with those from mice on a calorically restricted diet.

One interesting fact to emerge from the normal mouse cells was how few genes changed their pattern of activity as the mice aged. In apparent contradiction to the evolutionary theory of aging, which holds that large numbers of genes must be involved in aging, less than 1 percent of the tested genes became significantly more active in the muscles of the senescent mice, then thirty months old, and less than 1 percent became noticeably less active.

The genes that became more active in the aging mice were ones concerned with tissue repair; the less active genes included some involved in energy metabolism.

But all these changes were either absent or greatly reduced in thirty-month-old mice that had been on a calorically restricted diet. "Thus, at the molecular level, the CR [calorically restricted] mice appear to be biologically younger than animals receiving the control diet," Weindruch and Prolla concluded.[20]

The gene expression chip, they believe, provides for the first time a way to measure the aging of tissues at the cellular level. A chip programmed with human genes would enable a person's biological age, as distinct from their chronological age, to be measured. "It's our goal to test a patient's biological age from a drop of blood," Weindruch says.[21]

He and Prolla believe the pattern of genetic changes they saw in the normally fed and calorie restricted mice is consistent with a leading idea about the mechanism of aging, which has to do with the damage caused by energy metabolism.

The cell's main energy source is glucose, a sugar molecule whose energy is tapped by stripping off its hydrogen atoms, which are later combined with oxygen to make water. Oxygen, despite its familiarity, is a dangerous chemical, especially when in an active state. It links onto other molecules in the cell, forming chemically active substances known as free radicals. The free radicals crash around the cell like a bull in a china shop and attack working proteins or DNA. This mayhem is termed oxidative damage and is an inevitable side effect of extracting energy from glucose.

One theory about the mechanism of aging is that senescence occurs because of the slowly accumulating havoc from oxidative damage. Caloric restriction, the theory holds, extends life because less glucose is being burned in the body's cells and the burden of oxidative damage is therefore lighter. The changes in gene expression seen by Weindruch and Prolla were generally compatible with the idea that cells in normally aging mice face a greater burden of oxidative damage than do cells in dieting mice.

Genetic Manipulation of Life Span

A quite different approach to the study of aging is to manipulate the genes of laboratory organisms, particularly the usual suspects of roundworms, fruit flies, and mice. The extensions of life span achieved this way are more dramatic than those seen with caloric restriction. Several genes can be altered so as to double the usual life span of the roundworm, and a certain double gene mutation extends its lifetime sixfold, although the worms in this case are rather sluggish.

For the sake of comparison, a sixfold increase in human life span would give people in developed countries a life expectancy of 480 years and a maximum life span of 720 years, just 249 short of Methuselah's record.

The genetic manipulations are at present hard to reconcile with the evolutionary biologists' theory of aging. This powerful theory, as noted above, assumes that almost all the body's genes are optimized for a vigorous and healthy youth, leading to successful reproduction, but that natural selection has progressively diminished the ability to fine-tune healthy genes after the age of reproduction and parental care. The theory therefore predicts that no single gene can greatly affect life span because almost every gene gets increasingly out of whack as an individual's life progresses.

A single gene reduction in the rate of aging "contrasts interestingly with the predictions of evolutionary theory," says Linda Partridge.

Also, the idea that you can get something for nothing—just plug in a gene and double your life span—offends the trade-off concept that is implicit in the evolutionary theory. Partridge believes that either the experimenters failed to notice declines in the fertility of the longer-lived animals or the extra life span occurs only in the special conditions of the laboratory and not in the wild where the animal evolved.[22]

"People who think they are going to find any kind of fountain of youth, whether at the molecular or any other level, are not going to be successful," says the evolutionary biologist George Williams. "You have to look at the trade-offs. The reason a gene may prolong life is because it does other things that are deleterious. Evolution has optimized the distribution of genes over the life span. If there is some rare gene that retards senescence, it will have other effects that are not worth the benefit. If you starve mice, they live longer but don't reproduce."[23]

But some biologists who work on the roundworm believe, in almost complete defiance of the evolutionists' theory, that the worm's life span is under direct genetic control. Just as puberty is switched on in people at a particular stage in the life cycle, senescence in the worm is part of the life cycle program under genetic control. In the case of the worm, these biologists suspect, a hormone that promotes youthfulness is eventually switched off, and the worm quickly enters a terminal decline.

"It is often assumed that aging is the result of an inevitable process of decay and degeneration. However, there are reasons to believe that the aging process is actively regulated," writes Cynthia Kenyon of the University of California, San Francisco.[24]

Another biologist who studies the roundworm, Gary Ruvkun of the Harvard Medical School, believes that the worm's death is a genetically programmed event and that he has located the trigger of senescence in cells of the worm's brain. The mechanism involves insulin and energy metabolism, such fundamental features of animal cells that Ruvkun believes all animals, people included, may control life span in the same way. "Puberty and menopause come out of the central nervous system," he says, "and we believe that so too does when do you decide to die."

Should this be the case, the consequences would be considerable because human senescence could perhaps be controlled hormonally, just like the symptoms of menopause.

Embedded in the outer membrane of the worm's cells are receptor proteins that respond to the worm's equivalent of insulin. When the receptor detects insulin, it passes a signal to the cell's nucleus that switches on genes for the enzymes that metabolize glucose. This is how insulin controls the worm's rate of energy production.

Biologists can make roundworms live two to three times longer than usual by disrupting either the gene that makes the insulin receptor or another gene that mediates the signal from the receptor to the nucleus. The worms live longer, it's believed, because their cells, made deaf to insulin, burn glucose much more slowly, create fewer free radicals, and thus cause less oxidative damage to their working parts. This is the same oxidative damage theory that is invoked to explain why mice on calorically restricted diets live longer.

In support of the oxidative damage theory, roundworms have a special enzyme known as a catalase whose duty is to mop up free radicals. The catalase gene is switched on when roundworm larvae sense a shortage of food (or too many other roundworm larvae) and enter into a long-lived semi-hibernation mode. It seems that the catalase protects their cells from oxidative damage during their long hibernation phase. Martin Chalfie, a worm biologist at Columbia University, has found that if the catalase gene is disrupted, the hibernating worms do not live any longer than normal worms.[25] And researchers using a proprietary chemical that mimics catalase found that it increased the worms' life span by 44 percent, the first time life span has been enhanced by a chemical rather than by manipulating a gene.[26]

Ruvkun developed the interesting idea that oxidative damage could not be the general mechanism of senescence in the body, whether of worms, mice or people, because if it were, the tissues of the body would all give out at different times, depending on how much wear and tear they had had during life, whereas in fact they all seem to run down at about the same time. So could the oxidative damage be triggering senescence in a more centralized way, perhaps by damaging nerve cells that secrete a youthfulness hormone?

Roundworms are wonderful subjects for studying life span because their natural life span is less than three weeks. Mice may be more obviously like humans, but they live two years in the laboratory, so a life span extension study must last even longer. It's also very easy to manipulate the worms' genes because of the groundwork laid by worm geneticists and, since 1998, the availability of the worm's genome. Ruvkun took a long-lived strain of worms whose insulin receptor gene had been silenced and, by a clever genetic trick, turned the gene back on in specific tissues. Some worms had the gene reactivated in their muscles, some in their gut, and some in their nerve cells. It was only the worms in whose nerve cells the insulin receptor gene had been reactivated that reverted to the normal and shorter life span.[27]

The finding confirms the idea that the nerve cells are producing some factor that keeps the worms young and that preventing oxidative damage in these specific cells is what extends the worm's lifetime when the insulin receptor gene is disrupted. But proof of the idea will require the supposed youthfulness hormone to be isolated and tested.

Another fascinating attempt to pry apart the genetic mechanism of aging is being made across town from Ruvkun's laboratory by Leonard Guarente of the Massachusetts Institute of Technology. Guarente in 1991 decided to tackle the genetic basis of aging in yeast, a standard laboratory organism in which many basic aspects of living cells have been worked out. To study genes that affected longevity, he first needed to find longer lived strains of yeast.

After four years' work, he found a long lived yeast strain. Further analysis showed that its longevity was mediated by a gene whose protein silences other genes. When he inserted extra copies of the silencing gene into normal yeast strains, they lived longer, and when he disrupted their silencing gene, they died prematurely.

Keeping genes silenced is an essential activity in cells, since chaos would ensue if all the genes in the genome were switched on. But how might it be linked to aging? In studying the silencing protein's mode of operation, Guarente discovered accidentally that it needs an important chemical that is also used by the cell as an intermediate in glucose metabolism. This at once suggested a possible link with the phenomenon of caloric restriction and longevity. It seems that when cells are breaking down glucose, their stocks of the intermediate chemical are fully committed and the silencing protein cannot operate much. But if glucose is scarce, as in calorically restricted diets, the intermediate is freed up for use by the silencing protein, which then exerts its longevity effect.

Testing this idea, Guarente found that yeast cells starved of glucose lived longer, just like calorically restricted mice, but that the effect was abolished when their silencing gene was disrupted, showing that the longevity effect is mediated through the silencing gene.[28]

Worms, mice and people all possess genes that are related to the yeast silencing gene. Guarente is working his way up the evolutionary scale to see if the silencing gene is the mechanism through which the longevity/caloric restriction trade-off is made in these species too. He has found that worms with an extra copy of the silencing gene live longer, just as yeast cells do, and that the silencing gene seems to work through the same genetic pathway that extends life span when the insulin receptor gene is disrupted.[29]

The importance of research like Guarente's is that if the genetic mechanism that extends life span under caloric restriction can be identified, the genes involved can be targeted with drugs. Instead of trying to eat 30 percent fewer calories than usual, an impossible target for most people, one could perhaps just take a pill and get the same benefit.

A frequent theme in mythology is the punishment inflicted on those who dare to blur the line between mortals and gods by seeking eternal life. Another popular motif is of a golden age in which people lived forever in perfect health until some terrible act of impiety moved the gods to revoke their blessed state of immortality.

But the framers of such myths were perhaps seeking more to make death acceptable than to prohibit the search for longer life. For the first time in history, biologists have rational grounds for thinking it may be possible to extend human life span. "I believe our generation is the first to be able to map a possible route to individual immortality," says Haseltine.[30] Evolutionary biologists say that human life span is not fixed and immutable, though they doubt that in practice much can be done about it. Molecular biologists, knowing how similar their laboratory organisms are to people in many fundamental respects, hope to uncover the genetic mechanisms that will make human life extensible too.

At present their efforts are focused mostly on the life-extending powers of calorically restricted diets. Caloric restriction does not prevent death, only postpone it. But since the diet both extends life and offers protection against degenerative diseases, any drug or treatment that mimics the effects of caloric restriction would be well worth having.

Beyond caloric restriction, is immortality possible? Can a biological machine be patched up and kept running forever, like a car that is kept going indefinitely by replacing each part as it ages? We already possess a kind of immortality, though it applies only to our germ cells. Children are born equally young, whatever the age of their parents, because in some way the body makes time stand still for egg and sperm cells. Sooner or later, biologists will understand how this near miracle is performed and whether and at what cost it could be made more widely applicable.

7
Bravo, New World

Ah Love! could thou and I with Fate conspire
To grasp this sorry Scheme of Things entire
  Would not we shatter it to bits—and then
Re-mould it nearer to the Heart's Desire!

The human genome is a vast repository of infinitely valuable information. It contains the solutions, perfected by evolution over billions of years, for the optimum shape of all the working parts and genetic control regions of the human organism. What uses are likely to be made of this infinitely precious knowledge in the years and decades ahead?

The forecasting of technology is not a reliable art. It requires prediction not merely of technical progress—that's the easy part—but also of how society will pick and choose among the available options.

But the logic of developing the information in the genome is so compelling that its broad outlines seem clear enough. People desire health and long life for both themselves and their children, and knowledge from the genome will provide it. The question is whether the era of genome-based knowledge will be just another chapter in the history of medicine or a departure so radical as to transform human societies.

Genome technology, if it is allowed to follow its natural course, may be expected to develop in three overlapping waves of innovation which can be called conventional, germline, and life-extending.

The present intent is to use the human genome in the same way as all other medical research knowledge is used—to develop new and better treatments for disease. The impact of these conventional applications of genomic knowledge will be felt almost immediately in a minor way, and then with growing impact. As recounted in the previous pages, the first wave of drugs will be based directly on the body's own proteins, using either the proteins themselves or agents that inactivate them. Further ahead is another broad set of treatments based on knowledge of how gene variants contribute to specific diseases. Discovery of these gene variants, particularly those involved in multigenic diseases such as diabetes, heart disease and cancer, was a principal impetus for sequencing the human genome.

Now that researchers and physicians have access to the genome, the body's operating manual, and seem on the verge of being able to rebuild tissues with the body's stem cells, no insurmountable obstacle may remain to deriving rational treatments or cures for most human maladies. There may be serious exceptions to this general prediction, depending on how easy it is to reassemble the body's tissues from scratch. The brain, in particular, despite recent discoveries about its unexpected plasticity, may not be infinitely rewireable. Inherited mental diseases may be hard to reverse. So too may other conditions that result from a deep derangement of the body's construction as an embryo.

It is hard to imagine how to repair some of the body's more delicate feats of biological engineering, such as the eye or ear. The eye is constructed in the embryo from the delicate interaction between two different sheets of cells, neither of which exists as such in the adult. Still, the magic of regenerative medicine is that the physician does not have to know everything, only how to create the right conditions for the body's cells to respond to the appropriate signals. The information for making an eye or pancreas or kidney must exist within the genome, and can perhaps be re-evoked from it.

If the genome era should culminate in the happy closure of Pandora's box, with every human ailment confined and harmless, the ultimate goal of conventional medicine would have been achieved. People would enjoy good health to the end of their days and, save for those killed in accidents, most would live out their full life span. At present the maximum known human life span is 120 years. Probably few people have the genetic endowment to attain such a maximum. But if regenerative medicine should allow major organ systems to be patched or rejuvenated, more people might attain or even exceed the natural maximum, and the average human life span, which in developed countries has been steadily increasing over the last two hundred years, would perhaps continue its steady rise for many more decades.

The Era of Germline Genomics

The scenario sketched out above is merely the extrapolation of present trends. But just as likely are radical departures from current trends. These cannot be predicted but one obvious possibility, especially now that the genome sequence is available, is to make inheritable changes in the genome. There are many obvious dangers in altering the human germline, but consider for a moment the enormous advantages that genome germline engineering might offer, at least in principle.

Suppose it should become technically feasible to replace a disease-enhancing gene variant with the normal or health-enhancing version of the same gene. Such an intervention would doubtless be expensive for a single gene but much more cost effective if a large number of genes could be changed at the same time. And if such an intervention could save almost a lifetime's worth of medical treatment, it would be extremely cheap in light of the costs it would save.

The more expensive it becomes to treat an aging but longer lived population, whether with conventional methods or the tissue-rebuilding techniques of regenerative medicine, the cheaper the alternative of germline engineering is likely to appear.

But germline engineering would probably require a significant social change: a divorce between the procreational and recreational aspects of sex. Children could no longer be conceived the old-fashioned way if parents wished them to benefit from the best available genes in the human gene pool. Parents would probably need to go through the same procedures now used in fertility clinics, where oocytes are harvested from the woman's ovaries and fertilized in a dish with her partner's sperm.

This method would be necessary so as to allow genomic engineers access to the fertilized egg before it implants in the uterus. The egg would be treated with the corrective genes or exons that would replace or override the disease causing variants that either parent might contribute. Since one cell can be removed from an embryo without damage after the egg has reached the eight-cell stage, the treated embryos could be tested at this stage to see if the corrective genes had been properly incorporated.

Such methods of genetic engineering do not at present exist for human embryos. Somewhat equivalent approaches have been developed for mice, although mice embryos are used on a scale that would not be acceptable for humans.

Another approach might be to add a new pair of artificial chromosomes to the fertilized egg. Artificial chromosomes, now in the early stages of development, are DNA molecules equipped with sections that make the cell recognize and treat them as chromosomes. An artificial chromosome could carry a large package of corrective genes with ample room for DNA-based devices to override dominant disease-causing genes on other chromosomes.

Biologists can now manipulate DNA with considerable sophistication and will doubtless be able in future to build many interesting features into an artificial chromosome, such as an ability to receive updates either during or between generations.

Assuming that a method is invented for reengineering the human genome, it would be possible to go beyond the replacement of disease-causing gene variants and also insert genes that enhance desired qualities, such as strength, beauty, intelligence, and any other human attribute that has a genetic basis. A couple's children would be their own, but parents could select from among the most favorable gene variants in the human gene pool to enhance their children's qualities in ways they deem desirable.

Another advantage of germline genome engineering, besides alleviating much of the burden of disease, is that it could in principle reverse the unfairness of the genetic lottery. Good genes would be available by deliberate choice, not the accident of birth. The human genome is often said to be the patrimony of humankind, a phrase that would have precise meaning if everyone could indeed make use of it.

Genome Engineering: The Arguments Against

The possible benefits of genome engineering may be less familiar than the alleged disadvantages, many of which have been repeated for years in connection with gene therapy. There are indeed risks in genome engineering, as in any technology, and it will succeed only to the extent that the risks can be recognized and managed.

Gene therapy is the idea of introducing corrective genes into human cells, usually on the back of viruses because these agents have acquired through evolution the difficult trick of penetrating cells and integrating their genetic material into the chromosomes. Typically the gene therapist will remove the genes that make the virus cause illness and replace them with a therapeutic gene. Ethicists distinguish between gene therapy targeted at the ordinary cells of the body (somatic cell gene therapy) and gene therapy directed at the egg or sperm cells (germline gene therapy). So far only somatic cell gene therapy has been tried and despite twenty years of effort has proved largely unsuccessful.

There is general agreement that somatic cell gene therapy, should it work, is no different in principle from other surgical interventions that do not outlast the patient's own lifetime. But for ethical and other reasons it is considered premature to attempt germline gene therapy, and the subject is rarely discussed, in part because of the controversy it provokes. A recent exception was a 1998 conference on germline gene therapy at the University of California, Los Angeles.

The objections to genetically altering the human germline usually start with invocations of eugenics, a charge with relevance as a salutary though well worn reminder of past abuses. In practice, the central notion of eugenics has been to improve a whole population, usually defined as a race or nation, by sterilizing or killing those whose genes are judged undesirable by some government or institution.

Germline engineering, as currently discussed, does not aim at improving a race, and its methods will need to pass all ethical and legal requirements. The decisions would be made by parents trying to do what they deem is best for their children's health, not by governments. The merit of any technology depends on the social context in which it is used; eugenics refers to a context of the past that is unlikely to recur.

Another argument against germline change is that the technology is iniquitous because it would be more accessible to the rich than the poor. But it is hard to see why genetic engineering should have to bear the brunt of this effectively Luddite assertion. Almost all technologies, from the lightbulb to the microchip, have started out expensive and become cheaper as demand developed. To bar every new technology unless it is immediately as accessible as aspirin would be to bar all new technology.

Then it is argued that to change the germline is to play God. This contention is a slogan concealing a more serious argument. The slogan conveys the idea that the scope or power of genetic engineering is excessive by some standard known to the critic, but the standard is rarely articulated with clarity. The more interesting thought is that genetic engineering would substitute, in some perilous and adverse way, for the process that has shaped humans, which is evolution.

The impact of evolution on present day human societies is hard to measure and not well understood. To a large extent humans have long shaped their own environment, so that the effect of natural selection is less evident. The other motive force of evolution, a random change known as genetic drift, is less potent in the large societies of today than it may have been at the early stages of human existence. Evolution may not be taking us anywhere with the same speed as in the initial phases of human evolution. But mutation, on which natural selection works, is no less active than before. And most mutations are either neutral or harmful to the individuals in whose genomes they occur. It is not clear how much germline engineering would have to occur to show up at all above the incessant background noise of mutation.

Evolution shaped humans and human nature, yet it is a blind and pitiless process. Consider this extempore comment, made at the 1998 UCLA conference on germline engineering by Michael Rose:

"But just as you have to acknowledge the power and creativity of evolution, you also have to acknowledge its complete indifference to us as individuals. Evolution is about the transmission of DNA sequences down through time. We're just incidental things that get in the way. We're like the foot soldier in World War I, and we're sent out of the trenches into the enemy machine guns, and we die in our millions. That's fine with evolution as long as our DNA gets into the next generation. . . . And, of course, for those poor individuals who are afflicted by genetic diseases . . . their afflictions are a concrete example of where evolution has to be firmly rejected. . . . I think we need to seek an appropriate balance between respect for and use of what evolution does and rejection of what evolution does."[1]

Evolution works by selecting among the variations created by mutation. That abstraction has a bleak reality: most of those variations are our burden of genetic disease. If the phrase playing god has a serious meaning, it is that by acting to reduce genetic disease in the germline we might also be reducing the amount of variation on which evolution works.

This is a particular danger for humans because there is less natural variability in our population than in that of most species. Modern humans evolved comparatively recently, maybe as little as 50,000 years ago, and shortly before then went through a critically small bottleneck of about ten thousand or so breeding individuals. This small population had a much reduced genetic variability, and there has not been much time since to have acquired a lot more, despite the vast increase in our numbers.

If everyone in the world started choosing the same set of genes for their children, turning humankind into one vast clone, society would become insufferably tedious, besides which we could incur some of the dangers of inbred populations, such as susceptibility to epidemics.

But outlining a potential danger of genome engineering does not mean it will be fulfilled. It may be that the full roster of genes in a genetic health package would affect only a small percentage of the 30,000 genes in the genome and would reduce the number of variants of even these genes only slightly.

Genome Engineering and Human Nature

The true dangers of genome engineering lie way beyond the reach of easy slogans or easy solutions. They lie in the question of what changes should be permitted, if any, other than those directly related to health.

Sooner or later, though it is hard to say exactly when, biologists will understand almost everything about the human organism and the role of each gene in shaping the mind and body. By picking and choosing among the existing gene variants and perhaps even generating a few new ones, we will be able to modulate almost any aspect, making the effects of each gene more or less intense.

This includes the genes that affect human behavior. It is unknown, and a matter of some political controversy, how much of human behavior is genetically directed. Probably very little is rigidly fixed, as an ant's behavior is when it scents a pheromone telling it to follow a trail or defend its nest. Our social lives are too complicated for a fixed pattern of behavior to have any great survival value. Evolution, which can determine behavior very precisely in species where it is useful, has probably set mostly loose rules for human behavior. The biologist Edward O. Wilson calls these epigenetic rules. What they wire into the mind, in his view, is not specific behavior but general preferences or a trend to favor certain behaviors.

All human societies have a taboo against incest, for example. The behavior cannot be rigidly determined, because cases of incest are far from unknown. But given the universality of the taboo, there may be some general rule favored by the genes, such as an aversion to choosing as marriage partners those with whom one was intimate in childhood.

Other "instinctual algorithms" that guide humans through their life cycle, in Wilson's view, may include altruism, patriotism, religion, status seeking, territorial expansion, and contract formation.

Whatever the hereditary set of rules that defines human nature, it must be written into the genome and will therefore be subject to manipulation by genome engineers. Wilson believes that the ability to change the genome will mark a new era in human history, one that he calls "volitional evolution." "Evolution," he writes, "including genetic progress in human nature and human capacity, will be from now on increasingly the domain of science and technology tempered by ethics and political choice. We have reached this point down a long road of travail and self-deception. Soon we must look deep within ourselves and decide what we wish to become. Our childhood having ended, we will hear the true voice of Mephistopheles."[2]

Wilson believes that human nature, with all its imperfections, is a balance that evolution has worked out over thousands of years. To change any element, however undesirable it might seem in itself, would risk destroying the balance. Adolescent violence, for example, may spring from the same epigenetic rule that guides explorers and mountain climbers, he suggests. Any attempt to change the apparent imperfections in human nature "would lead to the domestication of the human species—we would turn ourselves into lapdogs," he says.[3]

"Alter the emotions and epigenetic rules enough," he writes in his book *Consilience,* "and people might in some sense be 'better,' but they would no longer be human. Neutralize the elements of human nature in favor of pure rationality, and the result would be badly constructed, protein-based computers. Why should a species give up the defining core of its existence, built by millions of years of trial and error?"[4]

As long as genome engineering sticks to changing genes that affect health, it will be on reasonably safe ground. It is easy to agree that people have the right to health and that, if the means of germline genome engineering is accepted, parents should be allowed to choose the genetic alterations that they believe will best safeguard their children's freedom from disease.

But how much further beyond health should the engineers go? Parents would like to enhance certain qualities in their children as well as just eliminate disease. Some would like their children to be a little taller than average, or a little smarter, or even a little kinder and more generous. This may seem harmless enough, but it is a process with no clear limits. Ethicists see an important distinction between health and enhancement. Changing the genome to promote health, defined as the absence of disease, can be considered as merely restoring an individual to some normal state. The required amount of genetic change is finite, and there is no prospect of producing an individual with unusual traits or any quality that could be deemed a departure from human nature.

Enhancing particular qualities, however desirable in themselves, may risk upsetting some important balance that evolution, in its wisdom, has found preferable. A Princeton University neurobiologist, Joe Tsien, reported in 1999 that he had genetically engineered a smarter strain of mouse, demonstrating the remarkable fact that nature has apparently neglected to maximize the mouse's intelligence. The change he engineered was an enhancement of the nerve cells' ability to associate two events, which lies at the basis of memory. In discussing Dr. Tsien's finding, some experts felt that too much memory could be bad (one might fail to forget bad experiences, becoming excessively fearful) and that nature had probably set the mouse's memory and intelligence at the level best suited to its overall chances of survival.

But Tsien doubts the natural wisdom argument, noting that nature does many things that are not in individuals' best interests, such as letting people's bodies run down and die. He believes that improving people's intelligence, whether by drugs or genetic alteration, could enhance a whole society. "Civilization is based on our extraordinary human intelligence," he says. "That is why our society evolves and civilization evolves, and if there is a way to enhance intelligence then it may not be surprising to see a change in the evolution of society."[5]

In deciding whether to allow enhancement of human qualities, a distinction could perhaps be drawn between physical qualities and behavioral ones. People might be allowed to improve any physical quality in their children, such as height, strength or skin complexion, but not to tamper with any mental attribute, for fear of changing human nature.

But if parents make frequent use of genomic engineering, it is easy to imagine governments too trying to get into the act, either directly or by offering incentives to influence parents' choices. Almost anything a government might want—obedient citizens, a nation of warriors, a congregation of fanatics, a quota of good computer programmers?—would be an unmitigated disaster. It's possible that genomic engineering would be confined to health, with enhancement off limits, or that if some forms of enhancement were allowed, the choice would be vested exclusively in parents with the strongest possible prohibition against any involvement by the state.

The Genome and Longevity

With the human operating manual in hand in the form of the human genome, physicians will doubtless be able to sustain the current trends toward increasing longevity. This incremental change in average life span can be accommodated without severe disruption. But a sudden jump in life span, however welcome for the beneficiaries, might be less easy to handle. Caloric restriction, as mentioned in the previous chapter, can increase life span by 30 percent in laboratory animals. Should biologists find a simple drug that mimics the effects of semi-starvation, human populations might start to live a lot longer overnight.

Biologists can already lengthen the life of the roundworm fourfold by changing certain of its genes. If an equivalent increase were engineered into the human germline, the human life expectancy in the developed world would stretch to not 80 but 320 years, and the maximum life span could reach 480 years.

These are extraordinary numbers and, if ever attained, they would perhaps change human society more than any other kind of enhancement. Human life is couched around the cycle of birth, procreation and death. If people become effectively immortal, the cycle makes less sense. Immortality, in many mythologies, is the defining difference between gods and mortals and, even if the myths are incorrect, not an attribute to be assumed lightly. Would one dare do anything so risky as carouse, drive a car, hit the ski slopes, if three hundred years of life would be thereby imperiled? Yet if these immortals sought to avoid any degree of risk, how would they avoid terminal boredom?

People with a life expectancy of three hundred years would not behave like ordinary people and would perhaps be so different as to be not entirely human. Human societies are built around the idea of finality and might be seriously undermined, at least in their present form, by tremendous longevity. If very long life is attained, new social arrangements might be needed for those who elect to benefit from it, perhaps in the form of special laws regarding the transfer of wealth between generations, or even the requirement that the Methuselahs live apart, on some Island of the Blessed or—why not?—Mars.

The Impending Choice

Contemplating the obvious dangers of widespread genome engineering, many would argue that it is better to declare the human germline sacrosanct and preclude entry into the minefield in the first place. The Genesis account, after all, reports that man is made in the likeness of God, implying that it would be presumptuous to attempt improvements. Genetic change, even in vegetables, seems to arouse deep public apprehensions, as shown by the recent uproar over genetically modified crops despite the absence of any clear danger.

Yet a technology that could offer greatly improved health to future generations probably cannot be fenced off forever as if it were an impassable minefield. James Watson, who set up the public consortium that sequenced the genome, has long been a keen advocate of changing the germline to eliminate genetic disease. "We can talk principles for ever but what the public actually wants is not to be sick. And if we help them not be sick, they'll be on our side," says Watson in an impassioned defense of germline engineering.

"It seems obvious," he said recently in a free-ranging comment, "that germline therapy will be much more successful than somatic. . . . The biggest ethical problem we have is not using our knowledge. We could have these techniques on hand so that we could at least see that the children who are going to be born won't die of a new plague. It's common sense to try to develop it. . . . We should be honest and say that we shouldn't just accept things that are incurable. I just think, 'What would make someone else's life better?' And if we can help without too much risk, we've got to go ahead and not worry whether we're going to offend some fundamentalist from Tulsa, Oklahoma."[6]

Genomic germline engineering is a technology that will not succeed unless public demand for it overwhelms the many obvious objections and apprehensions. Watson, for one, believes there will be a strong demand if germline engineering is allowed to develop without heavy-handed government regulation.

A development that may spur that demand is the impending use of genome-scan medical testing. Genome scans are likely to make people much more aware of their genetic constitution and more eager to improve it, if not in themselves then at least in their children. As the genetic basis of more and more human qualities becomes understood, people may take an interest in the genome scans of their prospective marriage partners. Probably everyone carries several disease-promoting gene variants, and people may desire to eliminate these from their children's heredity.

If genome engineering is to be tried, there will be many opportunities for designing the technology in a way that reduces risk. For one thing, all changes could probably be made reversible, so that extra genes inserted into an embryo could be inactivated later in life if the genes were found to have adverse consequences. This would offset a principal argument against changing the germline, that any mistakes would be made in perpetuity. Or the genes, perhaps on an artificial chromosome, could be designed so as to be upgradeable in future generations as better gene packages were designed.

"Clearly, developing improved humans will create great social and political problems with respect to unimproved humans," the physicist Stephen Hawking noted dryly in a 1998 White House address. "I'm not advocating human genetic engineering as a good thing, I'm just saying that it is likely to happen in the next millennium, whether we want it or not. This is why I don't believe science fiction like *Star Trek* where people are essentially the same four hundred years in the future. I think the human race, and its DNA, will increase its complexity quite rapidly."[7]

It is not so surprising that Hawking, a distinguished physicist who has been disabled with motor neuron or Lou Gehrig's disease for most of his career, should contemplate a world in which the human body can be designed and changed at will. He views genetic engineering as the power to accelerate the choices that evolution takes thousands of years to make. The improvement of humans could of course create serious problems with respect to the unimproved. All the more reason to make genetic enrichment available to all who may want it, perhaps a problem not so different from securing universal access to health care.

The dawn of genomic germline engineering throws light on the least just aspect of human societies, one long ignored because nothing could be done about it: the inequality of heredity. It is a soothing fiction that men are born equal. Inheritance is a random process in which some people inherit their parents' best genes, some their worst. Inheritance is a true lottery in which there are winners and losers, a pitiless system that exists so that evolution can back the winners.

The sequencing of the human genome makes it possible to envisage for the first time the creation of a genetically more just society, one in which the most fundamental kind of wealth—the genes that confer health and fitness—would for the first time be accessible to all.

Like all great gifts, this one will come with a price. The price is that, in beginning to alter the human germline in the name of health, we will inevitably assume a broader control over it. Evolution having done its part, our childhood will have ended and we will indeed hear the voice of Mephistopheles. But Faust heard a good angel too; he just made the wrong choice. Having come so far and gained our first glimpse of the human genome sequence, will we really declare it too dangerous to handle? If that were our nature, we would never have risked leaving the familiar African savanna in which we arose. We can never return to its confines. We can never turn back.

Notes

Note: All Web site URLs are subject to change.

1. THE MOST WONDROUS MAP

1. "Remarks by the President, The Entire Human Genome Project," The Office of Science and Technology Policy, Washington, D.C., June 6, 2000.
2. Neil A. Holtzman and Theresa M. Marteau, "Will Genetics Revolutionize Medicine?" *The New England Journal of Medicine* 343 (July 13, 2000): 141–144.
3. "Biotechnology Executives Discuss the Impact of the Genetic Revolution," *The Wall Street Journal,* July 24, 2000, p. B10.
4. Interview with author, July 14, 2000.
5. Karen P. Steel and Corné J. Kros, "A Genetic Approach to Understanding Auditory Function," *Nature Genetics* 77 (February 2001): 143–149.
6. Eric S. Lander and Robert A. Weinberg, "Genomics: Journey to the Center of Biology," *Science* 287 (March 10, 2000): 1777–1782.
7. Talk at the American Society for Clinical Oncology, New York, November 16, 2000.
8. Consider the discovery of penicillin, the antibiotic that became available during the Second World War in time to save the lives of millions of soldiers who would otherwise have succumbed to infected wounds. The discovery required a highly improbable series of events, which in-

cluded (a) the accidental seeding of a plate of *Staphylococcus* bacteria in Fleming's laboratory with a rare strain of mold, *Penicillium notatum,* that happened to produce copious amounts of penicillin; (b) a spell of mild weather in August 1928 that allowed the mold but not the bacteria to grow; (c) a period of hotter weather that prompted the staphylococci to bloom except around the mold; (d) the plate's remaining unsubmerged when Fleming on return from vacation dunked his plates in a bowl of disinfectant; (e) a chance visit and inquiry about the staphylococcus project by a colleague, which prompted Fleming to retrieve the plate and notice the clear zone around the mold; (f) Fleming's archiving of the strain, despite his erroneous conclusion that it would not make a useful therapy, so that ten years later its miraculous properties were rediscovered by Ernst Chain and Howard Florey. This synopsis rests on the account given by Gwyn Macfarlane in his book *Alexander Fleming* (London: Chatto & Windus, 1984).

2. THE RACE FOR THE HUMAN GENOME

1. Nicholas Wade, "Double Landmarks for Watson: Helix and Genome," *The New York Times,* June 27, 2000, p. F5.
2. Nicholas Wade, "Impresario of the Genome Looks Back with Candor," *The New York Times,* April 7, 1998, p. F1.
3. Robert M. Cook-Deegan, *The Gene Wars* (New York: W. W. Norton, 1994), p. 79.
4. Jeff Lyon and Peter Gorner, *Altered Fates* (New York: W. W. Norton, 1995), p. 533.
5. Ibid.
6. Cook-Deegan, *The Gene Wars,* p. 163.
7. Interview with author, December 1998.
8. Interview with author, December 1998.
9. Nicholas Wade, "It's a Three-Legged Race to Decipher the Human Genome," *The New York Times,* June 23, 1998, p. F3.
10. R. Waterston and J. E. Sulston, "The Human Genome Project: Reaching the Finish Line," *Science* 282 (October 2, 1998): 53–54.
11. Cook-Deegan, *The Gene Wars,* p. 311.
12. Ibid., p. 315.
13. Gina Kolata, "Biologist's Speedy Gene Method Scares Peers but Gains Backer," *The New York Times,* July 28, 1992, p. C1.
14. Nicholas Wade, "Bacterium's Full Gene Makeup Is Decoded," *The New York Times,* May 26, 1995, p. A16.

15. James D. Watson, "A Personal View of the Project," in *The Code of Codes,* ed. Daniel J. Kevles and Leroy Hood (Cambridge, Mass.: Harvard University Press, 1992), p. 164.

16. Nicholas Wade, "First Sequencing of a Cell's DNA Defines Basis of Life," *The New York Times,* August 1, 1995, p. C1.

17. Robert D. Fleischmann et al., "Whole-Genome Random Sequencing and Assembly of *Haemophilus influenzae* Rd," *Science* 269 (July 28, 1995): 496–512.

18. Nicholas Wade, "Genome Project Partners Go Their Separate Ways," *The New York Times,* June 24, 1997, p. C2.

19. Nicholas Wade, "Cambridge Lab Keeps Britain Ahead in Genome Stakes," *The New York Times,* October 6, 1998, p. F3.

20. An account of the complicated development of the ABI sequencer is given in Cook-Deegan, *The Gene Wars,* pp. 64–72.

21. Interview with author, May 9, 1998.

22. Ibid.

23. Ibid.

24. Ibid.

25. Richard Preston, "The Genome Warrior," *The New Yorker,* June 12, 2000, pp. 66–83.

26. Nicholas Wade, "In Genome Race, Government Vows to Move Up Finish," *The New York Times,* September 15, 1998, p. F3.

27. Nicholas Wade, "One of Two Teams in Genome Race Sets an Earlier Deadline," *The New York Times,* March 16, 1999, p. A21.

28. Nicholas Wade, "The Genome's Combative Entrepreneur," *The New York Times,* May 18, 1999, p. F1.

29. Nicholas Wade, "Gains Are Reported in Decoding Genome," *The New York Times,* May 22, 1999, p. A11.

30. Nicholas Wade, "Rivals Reach Milestones in Genome Race," *The New York Times,* October 26, 1999, p. F3.

31. *Genome Analysis,* vol. 3, *Cloning Systems,* ed. Bruce Birren et al. (Plainview, N.Y.: Cold Spring Harbor Laboratory Press, 1999), p. 6.

32. Nicholas Wade, "Company Nears Last Leg of Genome Project," *The New York Times,* January 11, 2000, p. F3.

33. Nicholas Wade, "On Road to Human Genome, a Milestone in the Fly," *The New York Times,* March 24, 2000, p. A1.

34. Nicholas Wade, "More Progress Reported in Decoding Key of Hereditary Information," *The New York Times,* March 30, 2000, p. A22.

35. Celera's report on its human genome sequence in *Science,* February 23, 2001, states, "The first assembly was completed June 25, 2000."

36. Maynard V. Olson, Testimony Before the House Committee on Science, Subcommittee on Energy and the Environment, June 17, 1998.
37. T. Friend, "The Book of Life: Twin Efforts Will Attempt to Write It," *USA Today,* June 9, 1998.
38. Nicholas Wade, "Rivals in the Race to Decode Human DNA Agree to Cooperate," *The New York Times,* June 22, 2000, p. A20.
39. Nicholas Wade, "Race Is On to Decode Genome of Mouse," *The New York Times,* October 6, 2000, p. A22.

3. The Meaning of the Life Script

1. Nicholas Wade, "Grad Student Becomes Gene Effort's Unlikely Hero," *The New York Times,* February 13, 2001, p. F1.
2. Genome sequences are stored in computers and viewed with special programs called browsers that reveal the genes and other known features. Kent's browser and assembled genome is available on the Web at http://genome.ucsc.edu.
3. Nicholas Wade, "Grad Student Becomes Gene Effort's Unlikely Hero."
4. Celera's article is: J. Craig Venter et al., "The Sequence of the Human Genome," *Science* 291 (February 16, 2001): 1304–1351. The consortium's is: The International Human Genome Sequencing Consortium, "Initial Sequencing and Analysis of the Human Genome," *Nature* 409 (February 15, 2001): 860–921.
5. Nicholas Wade, "Genome's Riddle: Few Genes, Much Complexity," *The New York Times,* February 13, 2001, p. F1.
6. Interview with author, March 22, 2000.
7. Interview with author, July 14, 2000.
8. Nicholas Wade, "Link Between Human Genes and Bacteria Is Hotly Debated by Rival Scientific Camps," *The New York Times,* May 18, 2001, p. A17.
9. A standard text, Bruce Alberts et al., *The Molecular Biology of the Cell,* 3d ed. (New York: Garland, 1994), pp. 1187–1189, lists 210 types of cell but does not attempt to categorize the neurons, or brain cells. According to a standard text on the brain, Eric R. Kandel et al., eds., *Principles of Neural Science,* 4th ed. (New York: McGraw-Hill, 2000), p. 67, more than 50 distinct types of neurons have been described. So at least 260 different types of human cell are already known, and the list seems certain to grow as expression chips allow

more careful distinctions to be made between the working anatomy of
different cell types.

10. Alberts et al., *The Molecular Biology of the Cell,* p. 337.

11. Talk at the American Society of Clinical Oncology, New York City,
November 16, 2000.

12. Scott Hensley, "Celera's Genome Anchors It Atop Biotech," *The Wall
Street Journal,* February 12, 2001.

13. E-mail to author, March 5, 2001.

14. Interview with author, April 11, 2001.

15. Interview with author, April 19, 2001.

16. E-mail to author, April 26, 2001.

17. Interview with author, April 16, 2001.

4. To Close Pandora's Box

1. David G. Wang et al., "Large-Scale Identification, Mapping and Genotyping of Single-Nucleotide Polymorphisms in the Human Genome,"
Science 280 (15 May 1998): 1077–1082.

2. Nicholas Wade, "Putting the Genome to Work," *The New York Times,*
June 27, 2000, p. F1.

3. Interview with author, August 22, 2000.

4. Interview with author, July 14, 2000.

5. David J. Weatherall, "Single Gene Disorders or Complex Traits:
Lessons from the Thalassaemias and other Monogenic Diseases,"
British Medical Journal 321 (November 2000): 1117–20.

6. Patrick O. Brown and David Botstein, "Exploring the New World of
the Genome with DNA Microarrays," *Nature Genetics* 21 (suppl.)
(January 1999): 33–37.

7. Richard A. Young, "Biomedical Discovery with DNA Arrays," *Cell*
102 (July 7, 2000): 9–15.

8. Ash A. Alizadeh et al., "Distinct Types of Diffuse Large B-Cell Lymphoma Identified by Gene Expression Profiling," *Nature* 403 (February 3, 2000): 503–511.

9. Tom Strachan and Andrew P. Read, *Human Molecular Genetics,* 2d
ed. (New York: Wiley-Liss, 1999), p. 398.

10. Sickle-cell anemia is caused by a SNP in the sixth codon of the gene
for the beta chain of hemoglobin, the oxygen-carrying protein of the
red blood cells. (A codon is a triplet of three successive nucleotides
that codes for an amino acid, the units of proteins. Hemoglobin con-

sists of four chains, two alpha chains and two beta chains, twisted together to form an elegant gas-transporting machine.)

The sixth codon of the hemoglobin beta chain in most people is GAG, which tells the ribosomes that the amino acid known as glutamic acid should be incorporated at that point in the protein chain. In people with sickle-cell anemia the codon's A is mutated to T. This atomic-level change has a bodywide effect. The codon GTG means that valine, not glutamic acid, is built into the hemoglobin chain. The switch of amino acids alters the physical properties of the hemoglobin molecules, making them more likely to clump together within the red blood cell instead of repelling one another and staying separate.

The clumping of hemoglobin distorts the red blood cells into a sickle form, in which shape they have trouble slipping through the smallest blood vessels. A traffic jam can form, blocking the vessels, depriving the tissues of oxygen, and leading to a serious crisis for the patient.

11. Yukio Horokawa et al., "Genetic Variation in the Gene Encoding Calpain-10 Is Associated with Type 2 Diabetes Mellitus," *Nature Genetics* 26 (October 2000): 163–175.

12. Interview with author, September 26, 2000.

13. Agnar Helgason et al., "Estimating Scandinavian and Gaelic Ancestry in the Male Settlers of Iceland," *The American Journal of Human Genetics* 67 (September 2000): 697–717.

14. Agnar Helgason et al., "mtDNA and the Origin of the Icelanders: Deciphering Signals of Recent Population History," *The American Journal of Human Genetics,* 66 (March 2000): 999–1016.

15. Michael Specter, "Decoding Iceland," *The New Yorker,* January 18, 1999, pp. 39–51.

16. Ricki Lewis, "Iceland's Public Supports Database, but Scientists Object," *The Scientist,* July 19, 1999, p. 1.

17. www.decodegenetics.com.

18. Vanessa Fuhrmann, "Roche Holding, deCODE to Unveil Pact," *The Wall Street Journal,* March 6, 2001.

19. J. Lazarou et al., "Incidence of Adverse Drug Reactions in Hospitalized Patients: A Meta-analysis of Prospective Studies," *The Journal of the American Medical Association* 279 (1998): 1200–1205.

20. Allen D. Roses, "Pharmacogenetics and the Practice of Medicine," *Nature* 405 (June 15, 2000): 857–865.

21. Wolfgang Sadee, "Pharmacogenomics," *British Medical Journal* 319 (November 13, 1999): 1–4.

22. C. Roland Wolf, Gillian Smith, and Robert J. Smith, "Pharmacogenet-ics," *British Medical Journal* 320 (April 8, 2000): 987–990.

23. William E. Evans and Mary V. Relling, "Pharmacogenomics: Translating Functional Genomics into Rational Therapeutics," *Science* 286, 487–491, October 15, 1999.

24. Connie M. Drysdale et al., "Complex Promoter and Coding Region β_2-Adrenergic Receptor Haplotypes Alter Receptor Expression and Predict *in vivo* Responsiveness," *Proceedings of the National Academy of Sciences* 97 (September 12, 2000): 10483–10488.

25. Scott Gottlieb, "Personalised Medicine Comes a Step Closer for Asthma," *British Medical Journal* 321 (September 23, 2000): 724.

26. Douglas Hanahan and Robert A. Weinberg, "The Hallmarks of Cancer," *Cell* 100 (January 7, 2000): 57–70.

27. Nicholas Wade, "Swift Approval for a New Kind of Cancer Drug," *The New York Times,* May 11, 2001, p. A1.

28. Brian J. Druker and Nicholas B. Lydon, "Lessons Learned from the Development of an Abl Tyrosine Kinase Inhibitor for Chronic Myelogenous Leukemia," *The Journal of Clinical Investigation* 105 (January 2000): 3–7.

29. Bert Vogelstein, David Lane, and Arnold J. Levine, "Surfing the p53 Network," *Nature* 408 (November 16, 2000): 307–310.

30. Nicholas Wade, "Treatment for Cancer Advances in Trials," *The New York Times,* August 1, 2000, p. F7.

31. Edwin A. Clark et al., "Genomic Analysis of Metastasis Reveals an Essential Role for RhoC," *Nature* 406 (August 3, 2000): 532–535.

32. T. R. Golub et al., "Molecular Classification of Cancer: Class Discovery and Class Prediction by Gene Expression Monitoring," *Science* 286 (October 15, 1999): 531–537.

33. Alizadeh et al., "Distinct Types of Diffuse Large B-Cell Lymphoma."

34. Lance Liotta and Emanuel Petricoin, "Molecular Profiling of Human Cancer," *Nature Reviews/Genetics* 1 (October 2000): 48–56.

5. Regenerative Medicine

1. Thomas B. Okarma, "Prospects for Cellular Therapies in the Treatment of Chronic Disease," *Journal of Commercial Biotechnology* 6 (Spring 2000): 300–307.

2. Nicholas Wade, "Teaching the Body to Heal Itself: Work on Cells' Sig-

nals Fosters Talk of a New Medicine," *The New York Times,* November 7, 2000, p. F1.

3. Michael G. Klug et al., "Genetically Selected Cardiomyocytes from Differentiating Embryonic Stem Cells from Stable Intracardiac Grafts," *The Journal of Clinical Investigation* 98 (July 1996): 216–224.

4. Nadya Lumelsky et al., "Differentiation of Embryonic Stem Cells to Insulin-Secreting Structures Similar to Pancreatic Islets," Sciencexpress, www.sciencexpress.org, April 26, 2001, due for later publication in *Science.*

5. Curt R. Freed et al., "Transplantation of Embryonic Dopamine Neurons for Severe Parkinson's Disease," *The New England Journal of Medicine* 344 (March 8, 2001): 710–719.

6. Irving L. Weissman, "Stem Cells: Units of Development, Units of Regeneration, and Units in Evolution," *Cell* 100 (January 7, 2000): 157–168.

7. Teruhiko Wakayama et al., "Differentiation of Embryonic Stem Cell Lines Generated from Adult Somatic Cells by Nuclear Transfer," *Science* 292 (April 27, 2001): 740–43.

8. Nicholas Wade, "Researchers Claim Embryonic Cell Mix of Human and Cow," *The New York Times,* November 12, 1998, p. A1.

9. Nicholas Wade, "Researchers Join in Effort on Cloning Repair Tissue," *The New York Times,* May 5, 1999, p. A22.

10. Nicholas Wade, "Panel Supports Use of Stem Cells for Research," *The New York Times,* September 15, 1999, p. A18.

11. Nicholas Wade, "Embryo Cell Research: A Clash of Values," *The New York Times,* July 2, 1999, p. A13.

12. Nicholas Wade, "Brain May Grow New Cells Daily," *The New York Times,* October 15, 1999, p. A1.

13. Elaine Fuchs and Julia A. Segre, "Stem Cells: A New Lease on Life," *Cell* 100 (January 7, 2000): 143–155.

14. Donald Orlic et al., "Bone Marrow Cells Regenerate Infarcted Myocardium," *Nature* 420 (April 4, 2001): 701–5.

15. Mark Sussman, "Hearts and Bones," *Nature* 410 (April 5, 2001): 640.

16. A. A. Kochler et al., "Neovascularization of Ischemic Myocardium by Human Bone-Marrow-Derived Angioblasts Prevents Cardiomyocyte Apoptosis, Reduces Remodeling and Improves Cardiac Function," *Nature Medicine* 7 (April 2001): 430–36.

17. Interview with author, August 30, 2000.

18. Fuchs and Segre, "Stem Cells."

19. Mary Hynes and Arnon Rosenthal, "Embryonic Stem Cells Go Dopaminergic," *Neuron* 28 (October 2000): 11–14.
20. William A. Haseltine, "The Signals of Life—A New Frontier in Medicine," unpublished manuscript, 2000.
21. Interview with author, September 6, 2000.
22. Interview with author, September 11, 2000.

6. The Quest for Immortality

1. Nicholas Wade, "Arguments over Life and the Need for Death," *The New York Times,* March 7, 2000, p. F4.
2. John R. Wilmoth, "In Search of Limits," in *Between Zeus and the Salmon: The Biodemography of Longevity,* ed. Kenneth W. Wachter and Caleb E. Finch (Washington, D.C.: National Academy Press, 1997), p. 46.
3. Marilyn Chase, "Centenarians' Genes May Unlock the Secret of a Long, Robust Life," *The Wall Street Journal,* January 14, 2000.
4. "And They All Lived Happily Ever After," *The Economist,* February 7, 1998, p. 82.
5. Wilmoth, "In Search of Limits."
6. Jean-Marie Robine and Michel Allard, "The Oldest Human," *Science* 279 (March 20, 1998): 1834.
7. Wachter and Finch, *Between Zeus and the Salmon,* p. 4.
8. Linda Partridge, "Evolutionary Biology and Age-related Mortality," in *Between Zeus and the Salmon,* p. 92.
9. Interview with author, March 21, 1998.
10. Interview with author, October 6, 2000.
11. Interview with author, December 23, 1997.
12. Caleb E. Finch, *Longevity, Senescence, and the Genome* (Chicago: University of Chicago Press, 1990), p. 306.
13. Steven N. Austad, *Why We Age* (New York: John Wiley and Sons, 1997), pp. 109–117.
14. Ibid., p. 104.
15. Finch, *Longevity, Senescence, and the Genome,* p. 208.
16. Michael R. Rose, *Darwin's Spectre* (Princeton, N.J.: Princeton University Press, 1998), p. 132.
17. S. Jay Olshansky, Bruce A. Carnes, and Douglas Grahn, "Confronting the Boundaries of Human Longevity," *American Scientist* 86 (January 1998): 52–61.

18. Richard Weindruch, "Caloric Restriction and Aging," *Scientific American,* January 1996, pp. 2–8.

19. Nicholas Wade, "A Pill to Extend Life? Don't Dismiss the Notion Too Quickly," *The New York Times,* September 22, 2000, p. A20.

20. Cheol-Koo Lee, Roger G. Klopp, Richard Weindruch, and Tomas A. Prolla, "Gene Expression Profile of Aging and Its Retardation by Caloric Restriction," *Science* 285 (August 27, 1999): 1390–1393.

21. Nicholas Wade, "New Study Hints at Way to Prevent Aging," *The New York Times,* September 23, 1999, p. A16.

22. Partridge, "Evolutionary Biology and Age-Related Mortality," p. 91.

23. Interview with author, December 1997.

24. Cynthia Kenyon, "Environmental Factors and Gene Activities That Influence Life Span," in *C. Elegans II,* ed. Donald L. Riddle et al. (Plainview, N.Y.: Cold Spring Harbor Laboratory Press, 1997), p. 792.

25. James Taub et al., "A Cytosolic Catalase Is Needed to Extend Adult Lifespan in *C. elegans* daf-C and clk-1 Mutants," *Nature* 399 (May 13, 1999): 162–166.

26. Simon Melov et al., "Extension of Life-Span with Superoxide Dismutase/Catalase Mimetics," *Science* 289 (September 1, 2000): 1567–1569.

27. Catherine A. Wolkow, Koutarou D. Kimura, Ming-Sum Lee, and Gary Ruvkun, "Regulation of *C. elegans* Life-Span by Insulinlike Signaling in the Nervous System," *Science* 290 (October 6, 2000): 147–150; Nicholas Wade, "Scientists Say Aging May Result from Brain's Hormonal Signals," *The New York Times,* October 10, 2000, p. F10.

28. Su-Ju Lin, Pierre-Antoine Defossez, and Leonard Guarente, "Requirement of NAD and *SIR2* for Life-Span Extension by Calorie Restriction in *Saccharomyces cerevisiae,*" *Science* 290 (September 2, 2000): 2126–2128.

29. Heidi A. Tissenbaum and Leonard Guarente, "Increased Dosage of a *sir-2* Gene Extends Lifespan in *Caenorhabditis elegans,*" *Nature* 410 (March 2–8, 2001): 227–230.

30. Interview, The Motley Fool, March 8, 2001; http://www.fool.com.

7. BRAVO, NEW WORLD

1. Quoted in Gregory Stock and John Campbell, eds., *Engineering the Human Germline* (New York: Oxford University Press, 2000), p. 93.
2. Edward O. Wilson, *Consilience* (New York: Alfred A. Knopf, 1998), p. 303.
3. Nicholas Wade, "From Ants to Ethics: A Biologist Dreams of Unity of Knowledge," *The New York Times,* May 10, 1998, p. F1.
4. Wilson, *Consilience,* p. 303.
5. Nicholas Wade, "Smarter Mouse Is Created in Hope of Helping People," *The New York Times,* September 2, 1999, p. A1.
6. Stock and Campbell, *Engineering the Human Germline,* p. 79.
7. Stephen W. Hawking, "Science in the Next Millennium," White House address, March 6, 1998.

Bibliography

Biomedical research has started to move so fast in the last few years that most of the results described in this book can only be found in what is euphemistically called the scientific literature. Although there are thousands of scientific journals, important biological discoveries tend to be published in just a handful, notably *Science* and *Nature,* which are rival weekly journals; *Cell;* the *Proceedings of the U.S. National Academy of Sciences,* known as *PNAS;* and *Nature*'s monthly spin-off journals known as *Nature Genetics* and *Nature Medicine.*

But perusal of these journals is not for the fainthearted. They are intended for specialists, not the general reader. To attain precision, biologists use a specialized vocabulary and deliberately colorless language. Here's a typical sentence: "Neovascularization of ischemic myocardium by human bone marrow–derived angioblasts prevents cardiomyocyte apoptosis, reduces remodeling and improves cardiac function." This is just the headline of an article that reports—though perhaps you failed to guess—a finding of quite considerable significance. A translation into the vernacular might go as follows:

Hey guys, just listen to this!!! We've found a totally novel kind of cell in human bone marrow! It seems to be the parent cell that gives rise to the body's blood vessels, so we're calling it an angioblast [Greek for grower of blood vessels]. These cool angioblasts cause the growth of new blood vessels [neovascularization] in heart muscle [myocardium] suffering from oxygen deprivation [ischemia]. The new blood supply prevents the self-destruction [apoptosis] of the

heart muscle cells [cardiomyocytes] that occurs after a heart attack.
It also blocks the usual pathological thickening and scar formation
[remodeling] and makes the heart work better than usual after a heart
attack. We used human angioblasts, but all these experiments were of
course done in rats, not people. (That's so obvious, there was no need
to say "rat" in the headline, was there?)

As the translation may show, scientific discourse requires fewer words
and is more precise to those who understand the technical terms. The con-
cepts are not difficult, once explained, but the daunting technical vocabulary makes most articles incomprehensible to non-scientists.

Even if the technical terms were plain, scientific articles are written in a
deliberately terse and unemotive style that seeks to veil rather than high-
light the significance of the argument. The purpose is to stress the writer's
objectivity and to avoid provoking critics who will be drawn, as sharks to a
bleeding carcass, to any claim with a hint of excess.

Even scientists have trouble understanding their fellow specialists' language, so the editors of many leading journals have adopted the salutary
habit of commissioning commentaries on important articles so as to flag
their significance and explain to other scientists why they are important.
For instance *Nature Medicine,* the journal in which the angioblast article
appeared, published a commentary with the lucid title, "Helping the heart
to heal with stem cells," and a clear statement of its method and meaning.

These journal commentaries are several levels closer to ordinary En-
glish than the original articles, yet the non-biologist is likely to find many
of these expositions also hard going. Therefore it is hard to recommend the
primary literature to readers unless they have acquired considerable familiarity with biologists' arcane argot.

At the risk of journalistic self-advertisement, the least strenuous way of
following the progress of molecular biology may be through newspapers
such as *The Wall Street Journal, The Washington Post* and *The New York
Times. The Wall Street Journal,* aimed at business readers, pays careful attention to the biotechnology industry and as part of its coverage provides
lucid explanations of advances in genomics and other frontiers of biological research.

Newspaper articles are intended for the general public, however, and
cannot get too technical without losing readers' attention, nor can they provide the copious background information that is required to understand the
new biological findings in depth. Readers seeking a deeper level of explanation can turn to the secondary literature. There have been several good
books about the genome project, gene therapy and other biological ad-

vances, the most recent of which is *Genome* (New York: HarperCollins, 1999) by Matt Ridley. *Genome* is a well conducted tour through several aspects of human genetics, but despite its title it focusses on interesting aspects of individual genes rather than the genome as such.

Probably the best way to understand the intellectual structure of modern biology is to open a textbook. Though this may seem a miserable way of passing an afternoon, a good textbook presents information in a full and orderly way, and summarizes vast amounts of knowledge. The later editions of leading texts have been market tested on cohorts of students and many attain high standards of exposition and explanatory graphics. And a textbook read for interest is not nearly so hateful as one that must be studied for a test.

That said, *The Molecular Biology of the Cell,* 3rd edition (New York: Garland Publishing, 1994) by Bruce Alberts and other authors, is one of the best gateways to modern biology. Addressed to college students, it assumes little prior knowledge, yet in clear though concise language takes the reader close to the forefront of current research. Even though at present the latest edition is that of 1994, the book's themes are so well chosen that it does not seem particularly out of date.

A more recent textbook, though not a substitute for the thoroughness of the Alberts volume, is *Genomes* (New York: Wiley-Liss, 1999) by T. A. Brown. Although it covers much the same ground, it aims, as the title implies, to make the genome rather than genes its major theme.

A remarkably up-to-date and comprehensive work is *Human Molecular Genetics,* 2nd edition (New York: Wiley-Liss, 1999) by Tom Srachan and Andrew P. Read. Unlike the previous two books, it focusses on human biology and pays little attention to the smaller organisms, such as yeast, *C. elegans* and *Drosophila,* in which most biological discoveries are worked out first. The book is also quite technical throughout, and in its quest for comprehensiveness is so condensed that some passages require slow and careful reading.

Though *The Molecular Biology of the Cell* may be the best technical introduction to modern biology, many readers may prefer to start with a quite different kind of book, *The Eighth Day of Creation,* expanded edition (Plainview, N.Y.: Cold Spring Harbor Laboratory Press, 1996) by Horace Freeland Judson. This work, a serious and entrancing piece of journalism, is also the best available historical account of the golden age of molecular biology, the two decades from the early 1950s when Crick and Watson divined the structure of DNA. This is where the story of the genome really begins, even though the word itself is barely mentioned.

Judson's tale, or at least the thread that leads to the genome project, is continued in *The Gene Wars* (New York: W. W. Norton, 1994) by Robert M. Cook-Deegan. In a pleasantly breezy style that conveys much careful research, the book recounts the genesis and early history of the Human Genome Project. Both Judson and Cook-Deegan paint vivid sketches of the scientists involved and the issues at stake. Their books will be helpful to any reader wishing to know more of the background to the events described in chapter 2.

The subject matter of chapter 3 (interpretation of the genome), chapter 4 (applications of the genome to medicine) and chapter 5 (stem cells and signalling molecules) is still so recent that it resides almost entirely in the scientific literature. The first analyses of the human genome have been available only since February 2001, and the field of stem cells has been expanding too rapidly for any of its practitioners to have been able to catch their breath and write a new textbook.

Research into aging and longevity, on the other hand, the theme of chapter 6, has a rich secondary literature. Several of the principal researchers have been moved to publish popular books in the last few years. Through an accident of timing, though, some appeared before the recent advances in the molecular biology of aging, such as the breaking of the Hayflick limit. These works include *Why We Age* (New York: John Wiley and Sons, 1997) by Steven N. Austad, *Cheating Time* (New York: W. H. Freeman, 1996) by Roger Gosden, *Time of Our Lives* (New York: Oxford University Press, 1999) by Tom Kirkwood, *The Quest for Immortality* (New York: W. W. Norton, 2001) by S. Jay Olshansky and Bruce A. Carnes, and *A Means to an End* (New York: Oxford University Press, 1999) by William R. Clark. All are good, with Austad's book in particular meriting attention for its lively style and graceful exposition.

For more specialist reading, a standard work in the field is *Longevity, Senescence, and the Genome* (Chicago: University of Chicago Press, 1990) by Caleb E. Finch. *Between Zeus and the Salmon*, edited by Kenneth W. Wachter and Caleb E. Finch (Washington, D.C.: National Academy Press, 1997), is a stimulating set of essays on various aspects of longevity.

Index

About the Author

Nicholas Wade was born in 1942 in Aylesbury, England, and educated at Eton and at King's College, Cambridge. He received a B.A. degree in natural sciences in 1964.

He worked for *Nature,* a weekly scientific magazine based in London, from 1967 to 1971, becoming deputy editor and Washington correspondent. In 1971 he joined the news staff of *Science,* a weekly scientific journal published in Washington, and in 1982 became a member of editorial board of *The New York Times,* writing editorials on science, health, environment and military technology.

He was science editor of *The New York Times* from 1990 to 1996, and has been a science reporter at the *Times* since 1997.

He is the author of several books, including *The Ultimate Experiment* (Walker, 1977), *The Nobel Duel* (Doubleday, 1981), *Betrayers of the Truth* (Simon & Schuster, 1983, written with William J. Broad) and *A World Beyond Healing* (Norton, 1987).

Printed in the United States
By Bookmasters